T0155503

Security Policy in System-on-Chip Designs

Sandip Ray • Abhishek Basak • Swarup Bhunia

Security Policy in System-on-Chip Designs

Specification, Implementation and Verification

 Springer

Sandip Ray
University of Florida
Austin, TX, USA

Abhishek Basak
Intel Corporation
Hilsboro, OR, USA

Swarup Bhunia
Electrical and Computer Engineering
University of Florida, Larsen Hall 216
Gainesville, FL, USA

ISBN 978-3-030-06666-6   ISBN 978-3-319-93464-8   (eBook)
https://doi.org/10.1007/978-3-319-93464-8

© Springer International Publishing AG, part of Springer Nature 2019
Softcover re-print of the Hardcover 1st edition 2019
This work is subject to copyright. All rights are reserved by the Publisher, whether the whole or part of
the material is concerned, specifically the rights of translation, reprinting, reuse of illustrations, recitation,
broadcasting, reproduction on microfilms or in any other physical way, and transmission or information
storage and retrieval, electronic adaptation, computer software, or by similar or dissimilar methodology
now known or hereafter developed.
The use of general descriptive names, registered names, trademarks, service marks, etc. in this publication
does not imply, even in the absence of a specific statement, that such names are exempt from the relevant
protective laws and regulations and therefore free for general use.
The publisher, the authors, and the editors are safe to assume that the advice and information in this book
are believed to be true and accurate at the date of publication. Neither the publisher nor the authors or
the editors give a warranty, express or implied, with respect to the material contained herein or for any
errors or omissions that may have been made. The publisher remains neutral with regard to jurisdictional
claims in published maps and institutional affiliations.

This Springer imprint is published by the registered company Springer Nature Switzerland AG
The registered company address is: Gewerbestrasse 11, 6330 Cham, Switzerland

Preface

This book is about System-on-Chip (SoC) security policies. Modern SoC designs contain a myriad sensitive information or "assets," which must be protected from unauthorized, malicious access. Security policies are rules and constraints that enable this protection. Unfortunately, security policies themselves are significantly complex, difficult to implement, and often error-prone. Furthermore, in current practice they are architected, designed, implemented, and deployed in an ad hoc manner, depending primarily on deep human insight. As we go to an ecosystem with billions to trillions of connected computing devices—each with its own unique security requirement—it is getting unrealistic to sustain this trend and still ensure effective security.

This book explores an alternative, i.e., an architectural framework for systematic design and implementation of SoC security policies. The approach entails definition of a specific, dedicated hardware block (often referred to as an "intellectual property" or "IP") that serves as the security brain of the SoC design. This security brain has the system-level security requirements programmed into it and can communicate with other IPs in the SoC design to enforce the programmed policy. This book dives into various facets of this architecture, including questions on power, performance, verification, and infield policy upgrades.

The research documented in this book originally appeared as part of the second author's Ph.D. dissertation. The research owes its existence substantially to the authors' frustration at the current state of the practice in security assurance, particularly at the system level. The practice in security architecture and validation today is far too dependent on human insight. As we move to ever more complex designs and an environment in which devices implementing these designs proliferate in billions and trillions, such approaches will no longer be tenable. Note that the approaches described are still at the stage of a research concept: they are not meant to be a comprehensive, industry-ready solution to the SoC security problem. However, we hope that it can help security practitioners and architects *think* about the kind of issues that need to be accounted for to develop a robust solution to a complex problem.

The research here owes significantly to discussions with our colleagues at Intel Corporation and NXP Semiconductors Inc. and colleagues and students at the the Case Western Reserve University and University of Florida. Jason Fung, Manoj Sastry, Sava Krstic, and Jin Yang at Intel provided numerous insights into the state of the practice of SoC security architecture, design, and validation. Lawrence Case, Wen Chen, and Monica Farkash at NXP also provided interesting insights and in particular served as sounding boards for a variety of the approaches discussed. We also thank Atul Prasad Deb Nath, particularly for the help with some of the works in Chaps. 5 and 6.

This research has been funded in part by the National Science Foundation and the Semiconductor Research Corporation, and we are grateful for the support.

Finally, we thank our friends and family for the help and support, particularly during the time spent in writing this book.

Gainesville, FL, USA   Sandip Ray
Hilsboro, OR, USA   Abhishek Basak
Gainesville, FL, USA   Swarup Bhunia
April, 2018

Contents

Chapter 1
SoC Security Policies: The State of the Practice

1.1 Overview

We are living in a world surrounded by billions of computing systems, identifying, tracking, and analyzing some of our intimate personal information including health, sleep, location, network of friends, etc. The trend is towards even higher proliferation of such devices, with an estimated 75B smart, connected devices by 2020. These devices generate, process, and exchange a large amount of sensitive information and data (often collectively referred to as "security assets" or simply, "assets"). In addition to private end-user information, assets include security-critical parameters introduced during the system architecture, e.g., fuses, cryptographic and DRM keys, firmware execution flows, on-chip debug modes, etc. Malicious access to these assets can result in leakage of company trade secrets for device manufacturers or content providers, identity theft or privacy breach for end users, and even destruction of human life.

Security assurance of a modern computing device is a challenging activity. One key challenge is the sheer complexity of the design. Most modern computing systems are architected via a System-on-Chip (SoC) paradigm, viz., through a composition of predesigned hardware or software blocks (referred to as "intellectual properties" or "IPs") that interact through a network of on-chip communication fabrics. The IPs themselves are highly complex artifacts optimized for performance, power, and silicon overhead. Adding to the complexity are the communication protocols used in implementing complex system-level use cases. Finally, security assets are sprinkled at different IPs across the design, and access to the assets is governed by complex security policies. The policies are defined by system architects as well as different IP and SoC integration teams, and undergo refinement and modification throughout the system development. This makes it challenging to validate a system, develop architectures to provide built-in resilience against unauthorized access, or update security requirements, e.g., in response to changing customer needs.

© Springer International Publishing AG, part of Springer Nature 2019
S. Ray et al., *Security Policy in System-on-Chip Designs*,
https://doi.org/10.1007/978-3-319-93464-8_1

This book is about a systematic approach for designing, implementing, and validating SoC security policies. The approach we advocate here makes use of a centralized IP block that is in sole charge of implementing and enforcing SoC security policies. Throughout this book, we will see examples of how such an architecture facilitates streamlined implementation, analysis, and deployment of diverse security policies for practical SoC designs.

This goal of this chapter is to set the context and provide the relevant background for the rest of the book. We discuss various types of security policies that must be enforced in a modern SoC designs, the current industrial practice in policy implementations, and the gaps and limitations in the state of the practice.

1.2   Introduction to Security Policies

At a high level, the definition of security requirement for assets in an SoC design follows the well-known "CIA" paradigm, developed as part of information security research [26]. In this paradigm, accesses and updates to secure assets are subject to the following three requirements:

- **Confidentiality:** An asset cannot be accessed by an agent unless authorized to do so.
- **Integrity:** An asset can be mutated (e.g., the data in a secure memory location can be modified) only by an agent authorized to do so.
- **Availability:** An asset must be accessible to an agent that requires such access as part of correct system functionality.

Of course, mapping these high-level requirements to constraints on individual assets in a system is nontrivial. Security policies are defined to achieve this goal. They specify which agent can access a specific asset and under what conditions. Following are two examples of representative security policies. Note that while illustrative, these examples are made up and do not represent security policy of a specific company or system.

Example 1:   During boot time, data transmitted by the cryptographic engine cannot be observed by any IP in the SoC other than its intended target.
Example 2:   A programmable fuse containing a secure key can be updated during manufacturing but not after production.

Example 1 is an instance of confidentiality, while Example 2 is an instance of integrity; however, the policies are at a lower level of abstraction since they are intended to be translated to "actionable" information, e.g., architectural or design features. The above examples, albeit hypothetical, illustrate an important characteristic of security policies: the same agent may or may not be authorized access (or update) of the same security asset depending on (1) the phase of the execution (i.e., boot or normal), or (2) the phase of the design life cycle (i.e., manufacturing or production). These factors make security policies difficult to implement. Exacerbating the problem is the fact that there is typically no central documentation for security policies; documentation of policies can range from microarchitectural and system integration documents to informal presentations and conversations among architects, designers, and implementors. Finally, the implementation of a policy is an exercise in concurrency, with different components of the policy implemented in different IPs (in hardware, software, or firmware), that coordinate together to ensure adherence to the policy.

1.3   System-Level Security Policies

Unfortunately, security policies in a modern SoC design are themselves significantly complex, and developed in ad hoc manner based on customer requirements and product needs. Below, we provide a taxonomy of security policy classes. They are not complete but illustrate the diversity of policies employed.

Access Control This is the most common class of policies, and specifies how different agents in an SoC can access an asset at different points of the execution. Here, an "agent" can be a hardware or software component in any IP of the SoC. Examples 1 and 2 above are examples of such policy. Furthermore, access control forms the basis of many other policies, including information flow, integrity, and secure boot.

Information Flow Values of secure assets can sometimes be inferred without direct access, through indirect observation or "snooping" of intermediate computation or communications of IPs. Information flow policies restrict such indirect inference. An example information flow policy might be the following:

- *Key Obliviousness:* A low-security IP cannot infer the cryptographic keys by snooping only the data from crypto engine on a low-security communication fabric.

Information flow policies are difficult to analyze. They often involve highly sophisticated protection mechanisms and advanced mathematical arguments for correctness, typically involving hardness or complexity results from information security. Consequently, they are employed only on critical assets with very high confidentiality requirements.

Liveness These policies ensure that the system performs its functionality without "stagnation" throughout its execution. A typical liveness policy is that a request for a resource by an IP is followed by an eventual response or grant. Deviation from such a policy can result in system deadlock or livelock, consequently compromising system availability requirements.

Time-of-Check vs. Time-of-Use (TOCTOU) This refers to the requirement that any agent accessing a resource requiring authorization is indeed the agent that has been authorized. A critical example of TOCTOU requirement is in firmware update;

the policy requires firmware eventually installed on update is the same firmware that has been authenticated as legitimate by the security or crypto engine.

Secure Boot Booting a system entails communication of significant security assets, e.g., fuse configurations, access control priorities, cryptographic keys, firmware updates, debug and post-silicon observability information, etc. Consequently, boot imposes more stringent security requirements on IP internals and communications than normal execution. Individual policies during boot can be access control, information flow, and TOCTOU requirements; however, it is often convenient to coalesce them into a unified set of boot policies.

Note that the policies above relate to *integration* characteristics of SoC designs, not to fidelity of individual IPs. For now, we ignore the possibility that the IPs themselves might be untrustworthy, i.e., the threat model underlying the security policies includes external attacks through software or SoC interface but not malicious hardware introduced in the IPs themselves. This threat model is reasonable for SoC integration flows that involve primarily in-house rather than third-party IPs or in which IPs are integrated after (orthogonal) fidelity checks for existence of malicious backdoors or Trojans. We will consider untrusted IPs in Chap. 4.

1.4 Communication Policies

In addition to the system-level policies, there are also "lower-level" policies that govern architecture of IPs and their communication requirements. For example, communication along an NoC is specified by *fabric policies*. Following are some of the obvious fabric policies:

Message Immutability If IP $\mathscr{A}$ sends a message m to IP $\mathscr{B}$, then the message received by $\mathscr{B}$ must be exactly message m.

Redirection and Masquerade Prevention If $\mathscr{A}$ sends a message m to $\mathscr{B}$, then the message must be delivered to $\mathscr{B}$. In particular, it should be impossible for a (potentially rogue) IP $\mathscr{C}$ to masquerade as $\mathscr{B}$, or for the message to be redirected to a different IP $\mathscr{D}$ in addition to, or instead of $\mathscr{B}$.

Non-observability A private message from $\mathscr{A}$ to $\mathscr{B}$ must not be accessible to another IP during transit.

The descriptions of the above policies perhaps belie the complexity involved in implementing them. To understand some of the subtleties, consider the SoC configuration shown in Fig. 1.1. Suppose that IP IP0 needs to send a message to the DRAM. Ordinarily, the message would be routed through Router3, Router0, Router1, and Router2. However, such a route permits message redirection via software. In particular, each router includes a base address register (BAR) which

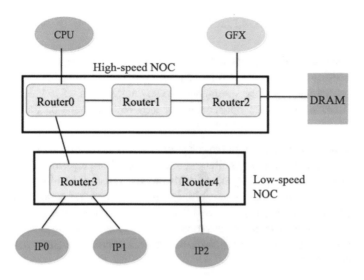

Fig. 1.1 An illustrative toy SoC configuration. Typical SoC designs include several on-chip fabrics with differing speed and power consumption profiles. For this toy configuration, we assume a high-speed fabric with three routers connected linearly, and a low-speed fabric with two routers also connected linearly

is used to route messages for specific destinations. However, one of the routers in the proposed path, Router0, is connected to the CPU; the BARs in this router are subject to potential overwrite by a potentially malicious host operating system, which can consequently redirect a message passing through Router0 to a different destination. Consequently, a secure message cannot be sent from IP0 through DRAM via this route unless the host operating system is trusted. Note that understanding the potential of redirection in this scenario requires knowledge of operation of the fabrics, functioning of routers within an NoC (e.g., the use of BARs) as well as the capabilities of the software in an adversarial role.

In addition to the above generic policies, most SoC designs typically include additional asset-specific communication constraints. For instance, a potential fabric policy relevant to secure boot is listed below. This policy ensures that a key generated by the fuse controller cannot be sniffed during propagation to the crypto engine for storage.

• **Boot-time key non-observability:** During the boot process, a key from the fuse controller to the crypto engine cannot be transmitted through a router to which any IP with user-level output interface is connected.

1.5   Security Along SoC Design Life Cycle

Figure 1.2 provides a high-level overview of the SoC design life cycle. Each component of the life cycle, of course, involves a large number of design, development, and validation activities. Here, we summarize the key activities involved along the life cycle that pertain to security. Subsequent sections will elaborate on the individual activities.

Risk Assessment Security requirements definition is a key part of product planning, and happens concurrently with (and in close collaboration with) the definition of architectural features of the product. This process involves identifying the security assets in the system, their ownership, and protection requirements, collectively defined as *security policies* (see below). The result of this process is typically the generation of a set of documents, often referred to as *product security specification* (PSS), which provides the requirements for downstream architecture, design, and validation activities.

Security Architecture The goal of a security architecture is to design mechanisms for protection of system assets as specified by the PSS. It includes several components, including (1) identifying and classifying potential adversary for each asset; (2) determining attacker entry points, also referred to as threat modeling; and (3) developing protection and mitigation strategies. The process can identify additional security policies—typically at a lower level than those identified during risk assessment (see below)—which are added to the PSS. The security definition typically proceeds in collaboration with architecture and design of other system

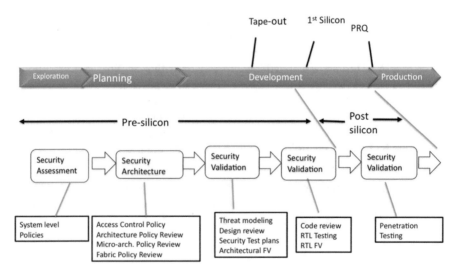

Fig. 1.2 SoC design life cycle

features, including speed, power management, thermal characteristics, etc., with each component potentially influencing the others.

Security Validation Security validation represents one of the longest and most critical parts of security assurance for industrial SoC designs, spanning the architecture, design, and post-silicon components of the system life cycle. The actual validation target and properties validated at any phase, of course, depends on the collateral available in that phase, e.g., we target, respectively, architecture, design, implementation, and silicon artifacts as the system development matures. Below, we will discuss some of the key validation activities and associated technology. One key component of security validation is to develop techniques to subvert the advertised security requirements of the system, and identify mitigation measures. Mitigation measures for early stage validation targeting architecture and early system design often includes significant refinement of the security architecture itself. At later stages of the system life cycle, when architectural changes are no longer feasible due to product maturity, mitigation measures can include software or firmware patches, product defeature, etc.

1.6 Elements of Security Architecture

Given a plethora of complex policies and protection requirements under different classes of potential adversaries, how would we go about designing authentication mechanisms to ensure policy enforcement? Unfortunately, the state of the practice in this area today depends heavily on human creativity and observation. The typical approach today is to develop a baseline architecture definition which is then repeatedly refined through the following two steps:

- Use threat modeling to identify potential threats to the current architecture definition (see below);
- Refine the architecture with mitigation strategies covering the threats identified.

The baseline architecture is typically derived from legacy architectures for previous products, adapted to account for the policies defined for the SoC design under exploration. In particular, for each asset, the architect must identify (1) who can access the asset, (2) what kind of access is permitted by the policies, and (3) at what points in the system execution or product development life cycle such access requests can be granted or denied. The process can be highly complex and tedious for several reasons. In particular, an SoC design may have a significant number of assets, often of the order of thousands if not more. Furthermore, not all assets are statically defined; many assets are created at different IPs during the system execution. For example, a fuse or e-wallet may have a statically defined asset such as key configuration modes. During system execution, these modes are passed to the cryptographic engine, which generates the cryptographic keys for different IPs and transmits them through the system NoC to the respective IPs. Each participant in this process has sensitive assets (either static or created) during different phases of the system execution, and the security architecture must account for any possible access to these assets at any point, possibly under the relevant adversary model.

1.7 Threat Modeling

Threat modeling is the name of the activity for optimizing SoC security by identifying objectives and vulnerabilities, and then defining countermeasures to prevent, or mitigate the effects of, threats to the system. As noted above, it is a vital part of the security architecture definition. It is also a key part of the security validation, in particular in negative testing and white-box hacking activities. Threat modeling roughly involves the following five steps, which are iterated until completion:

Asset Definition Identify the system assets governing protection. This requires identification of IPs and the point of system execution where the assets originate. As discussed above, this includes statically defined assets as well as those generated during system execution.

Policy Specification For each asset, identify the policies that involve it. Note that a policy may "involve" an asset without specifying direct access control for it. For example, a policy may specify how a secure key $\mathcal{K}$ can be accessed by a specific IP. This in turn may imply how the controller of the fuse where $\mathcal{K}$ is programmed can communicate with other IPs during boot process for key distribution.

Attack Surface Identification For each asset, identify potential adversarial actions that can subvert policies governing the asset. This requires identification, analysis, and documentation of each potential "entry point," i.e., any interface that transfers data relevant to the asset to an untrusted region. The entry point depends on the category of the potential adversary considered in the attack, e.g., a covert-channel adversary can make use of nonfunctional design characteristics such as power consumption or temperature to infer the ongoing computation.

Risk Assessment The potential for an adversary to subvert a security objective does not, in and of itself, warrant mitigation strategies. The risk assessment and analysis are defined in terms of the so-called DREAD paradigm, composed of the following five components: (a) Damage potential; (b) Reproducibility; (c) Exploitability, i.e., the skill and resource required by the adversary to perform the attack; (d) Affected systems, e.g., whether the attack can affect a single system or tens or millions; and (e) Discoverability. In addition to the attack itself, one needs to analyze the likelihood that the attack can occur on-field, motives of the adversary, etc.

Threat Mitigation Once the risk is considered substantial given the likelihood of the attack, protection mechanisms are defined and the analysis must be performed again on the modified system.

Implementation Example Consider protecting a system against code-injection attacks by malicious or rogue IPs by overwriting code segments through Direct Memory Access (DMA). The assets being considered here are appropriate regions of memory hierarchy (including cache, SRAM, and secondary storage), and the governing policy may be used to define DMA-protected regions where DMA access is disallowed. The security architect needs to go through all memory access points in the system execution, identify memory access requests to DMA-protected regions, and set up mechanisms so that DMA requests to all protected accesses will fail. Once this is done, the enhanced system must be evaluated for additional potential attacks, including attacks that can potentially exploit the newly setup protection mechanisms themselves. Such checks are performed typically via *negative testing*, i.e., looking beyond what is specified to identify if the underlying security requirements can be subverted. For our DMA protection example, such testing may involve looking for ways to access the DMA-protected memory regions, other than directly performing a DMA access. The process is iterative and highly creative, resulting in a collection of increasingly complex lineup of protection mechanisms, until the mitigation is considered sufficient with respect to the risk assessment.

Obviously, performing the above activities manually over the range of system assets and policies, in the presence of subtleties related to implicit expectations, potential adversaries that break the risk/mitigation analysis, and complex interplay between functional behavior and security constraints, is a daunting task. Admittedly, there is a host of available tools to assist in the different steps, e.g., tools for documenting steps in threat and severity identification identifying security scenarios, etc. [46, 67]. Nevertheless, the key architectural decisions and analysis still depend highly on human insights.

1.8   Security Validation Overview

Designing resilience into designs is one aspect of security assurance. The other critical aspect is validating that the security objectives of the product are indeed satisfied. In this section, we provide an overview of the different security validation objectives; the next section will discuss the technologies used.

Unfortunately, the role of security validation is different from most other kinds of validation (such as functional or power-performance or timing) since the requirements are typically less precise. In particular, the goal of security validation is to "validate conditions related to security and privacy of the system that are not covered by other validation activities." The requirement that security validation focuses on targets not covered by other validation is important given strict time-to-market constraints, which preclude duplication of resources for the same (or similar) validation tasks; however, it puts onus on the security validation organization to understand activities performed across the spectrum of the SoC design validation and identify holes that pertain to security. To exacerbate the problem, a significant amount of security objectives are not clearly specified, making it difficult to (1) identify validation tasks to be performed, and (2) develop clear coverage/success criteria for the validation. Consequently, the validation plan includes a large number of diverse activities that range from the science to the art and sometimes even "black magic."

At a high level, security validation activities can be divided roughly along the following four categories:

Functional Validation of Security-Sensitive Design Features This is essentially extension to functional validation, but pertains to design elements involved in critical security feature implementations. An example is the cryptographic engine IP. A critical functional requirement for the cryptographic engine is that it encrypts and decrypts data correctly for all modes. As with any other design block, the cryptographic engine is also a target of functional validation. However, given that it is a critical component of a number of security-critical design features, security validation planning may determine that correctness of cryptographic functionality to be crucial enough to justify further validation beyond the coverage provided by vanilla functional validation activities. Consequently, such an IP may undergo more rigorous testing, or even formal analysis in some cases. Other such critical IPs may include IPs involved in secure boot, on-field firmware patching, etc.

Validation of Deterministic Security Requirements Deterministic security requirements are validation objectives that can be directly derived from security policies. Such objectives typically encompass access control restrictions, address translations, etc. Consider an access control restriction that specifies a certain range of memory to be protected from DMA access; this may be done to ensure protection against code-injection attacks, or protect a key that is stored in such location, etc. An obvious derived validation objective is to ensure that all DMA calls for access to a memory whose address translates to an address in the protected range must be aborted. Note that validation of such properties may not be included as part of functional validation, since DMA access requests for DMA-protected addresses are unlikely to arise for "normal" test cases or usage scenarios.

Negative Testing Negative testing looks beyond the functional specification of designs to identify if security objectives can be subverted or are underspecified. Continuing with the DMA-protection example above, negative testing may extend the deterministic security requirement (i.e., abortion of DMA access for protected memory ranges) to identify if there are any other paths to protected memory in addition to address translation activated by a DMA access request, and if so, potential input stimulus to activate such paths.

Hackathons Hackathons, also referred to as *white box hacking* fall in the "black magic" end of the security validation spectrum. The idea is for expert hackers to perform goal-oriented attempts at breaking security objectives. This activity depends primarily on human creativity, although some guidelines exist on how to approach them (see discussion on penetration testing in the next section). Because of their cost and the need for high human expertise, they are performed for attacking complex security objectives, typically at hardware/firmware/software interfaces or at the chip boundary.

1.9 Summary

It is clear from the discussions so far that design, implementation, and validation of SoC security policies is a complex and chaotic activity encompassing the entire system development life cycle and reconciling needs and interests of a variety of stakeholders. Security assets in SoC designs spread across multiple IP blocks. Their access restrictions often involve subtle and complex interactions between hardware, firmware, and/or software associated with these design modules. Hence, security policies controlling operations involving these security assets are often complex and ambiguous, which makes it difficult for the SoC designers to correctly implement them. To exacerbate the issue, security policies are rarely specified in any formal, analyzable form. Some policies are described (in natural language) in different architecture documents, and many remain undocumented. Along with the increased complexity arising during design, potentially incurring greater resources and design time (affecting time-to-market), this current practice leads to the following major issues:

- It becomes extremely difficult to validate the SoC for adherence to the system security requirements during post-Si validation and/or in-field testing. In the absence of a formal, methodical approach, bugs or errors detected during security validation are becoming increasingly complicated to trace back to their sources and thus correct them. This potentially requires more validation time and resources, thereby leading to increased time-to-market. Indirectly, the probability of vulnerabilities remaining in an SoC design after design or test/patch is also enhanced.
- It is very complicated to patch/upgrade the SoC security policies which might be necessary in response to bugs found in-field as well as dynamically changing security requirements due to varying product/system usage scenarios, including its adoption in different geographic market segments around the world.
- The approach of systematic design reuse based on which modern SoCs are implemented is hampered (in context of secure SoC design) due to the ad hoc nature of implementation of security policies. This too often multiplies design effort and complexity and leads to increased time-to-market.

In the light of these pressing issues, there is a critical need for devising a systematic approach for design, implementation, and validation of SoC security policies. In the rest of the book, we will develop one such approach, based on a novel security architecture based on a centralized IP for security policy implementation.

The approach presented in this book has found application in realistic policy implementations, and we document some of the results in the various chapters. However, the results are to be taken with a caveat. Security needs and constraints vary based on the kind of product, the target for its deployment, etc. Consequently, the cost of security implementations also varies, and the results in this book may not be reflective of all the scenarios. Nevertheless, we hope that the analyses provided in this book will be still valuable as representative of the *kind* of parameters and trade-offs that a designer ought to make in order to apply a similar systematic approach for their target domain.

1.10   Bibliographic Notes

SoC security assurance has emerged as a critical research area in recent years. There are many excellent surveys and tutorials on various sub-topics in the area [12, 60, 74]. The notion of security policies has a long history starting with several seminal works in the 1980s [24, 26, 30, 62]. More recently, with the increasing prominence of SoC designs, there has been significant research interest in SoC security. However, most recent research in this area has focused on *hardware security*, i.e., protection of the system against a malicious hardware Trojan [68], detection of counterfeit IP [27], etc. Some recent papers by the authors [57, 58, 60] have touched upon aspects of industrial practice in SoC security architecture and validation.

Chapter 2
E-IIPS: A Centralized Policy Implementation Architecture

2.1 Introduction

This chapter will introduce E-IIPS (Extended Infrastructure IP for Security)—an architecture for systematic and flexible implementation of SoC security policies. The description of the architecture in this chapter is fairly basic. We introduce its key features and demonstrate how basic system-level policies are implemented through this architecture. We will progressively enhance the architecture with advanced features in the next few chapters.

The key idea of E-IIPS is simple. One of the key reasons why security architecture and validation is difficult in today's practice is that assets and their protection mechanisms are sprinkled across the numerous IPs in the SoC design. This makes it nontrivial to validate that the protection mechanisms indeed implement the required policies. It also makes it difficult to upgrade the policies in-field, possibly in response to changing security requirements. Consequently, one way to address this is to develop an architecture where there is a single point of implementation of all security policies. The idea of E-IIPS is a realization of this intuition. In the architectural framework implemented in E-IIPS, policies are all implemented within a single, centralized IP that serves as the security brain of the SoC design. This IP obviously needs to communicate with the other IPs in the SoC design to enforce the numerous system-level policies. However, the policies (and even the communications necessary to enforce them) are all implemented in a single, centralized IP. Consequently, verification and in-field upgrade activities need to focus only on this single IP, making it streamlined. Perhaps more importantly, it enables the architect to implement security policies systematically within a single, uniform architectural framework.

© Springer International Publishing AG, part of Springer Nature 2019
S. Ray et al., *Security Policy in System-on-Chip Designs*,
https://doi.org/10.1007/978-3-319-93464-8_2

2.2   Architecture

Figure 2.1 illustrates our proposed architecture. It includes two main components: (1) a centralized security policy controller IP (referred to as E-IIPS or extended IIPS in the rest of the paper), and (2) security wrappers around individual IPs to facilitate communication with E-IIPS. To facilitate configurability across different products and use cases, E-IIPS is defined as a microcontrolled soft IP. SoC designers can program security policies through E-IIPS as firmware modules that are then stored in a secure ROM or flash memory; secure policy update is supported through an authenticated firmware update mechanism. E-IIPS communicates with other IPs via corresponding security wrappers as follows. For enforcing different security policies, E-IIPS may need different local IP-specific collateral. For instance, suppose a policy prohibits access of internal registers of IP A by IP B when A is in the middle of a specific security-critical computation. To enforce the policy, E-IIPS must "know" when B attempts to access the local registers of A as well as the security state of the computation being performed by A. The security wrappers provide a standardized way for E-IIPS to obtain such collateral while abstracting the details of internal implementation of individual IPs. In particular, the wrappers implement a protocol to communicate with E-IIPS during the execution. Based on the policies implemented, E-IIPS can configure the wrapper of an IP at boot time to provide internal event information under specific conditions (e.g., security status of internal computation, read requests to specific IPs, etc.); the security wrappers monitor for configured conditions and provide requested notification to E-IIPS. IP development teams are responsible for augmenting individual IP with the security wrapper, by extracting security-critical information (see below).

Design Choices A key design choice for E-IIPS is its centralized firmware-upgradable architecture, i.e., it is implemented as a single reusable IP block in the SoC. This choice is governed by the need to provide a single place for understanding, exploration, upgrade, and validation of system-level security policies. Indeed, the current complexity in security policy analysis and modification is precise that the policies are "sprinkled" across the different IPs in the SoC. Our centralized architecture is specifically intended to alleviate this complexity. On the other hand, this choice implies that communication with E-IIPS is a bottleneck for system performance. We address this issue by making the security wrappers "smart" so that only security-relevant information is communicated to E-IIPS, possibly under the latter's directive. Finally, the choice of a microcontrolled rather than hardware implementation stems from the need to update security policies on-field, either due to customer requirements or in response to a known exploit or design bug. On the other hand, this makes E-IIPS itself vulnerable to attacks through firmware updates. In Sect. 2.2.4, we discuss authentication mechanisms to address this issue.

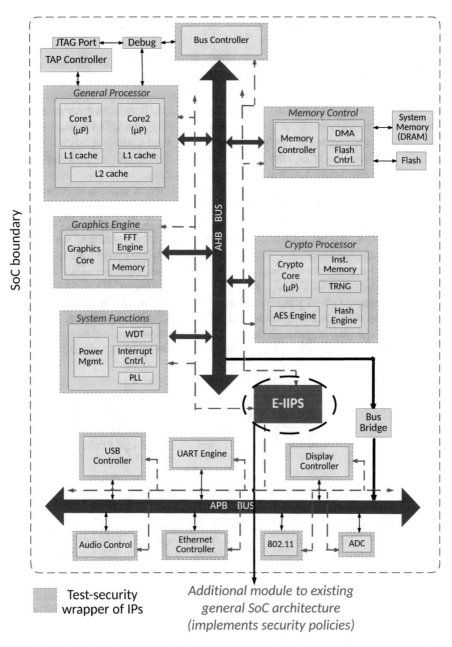

Fig. 2.1 Schematic of a proposed architecture framework with the major components, for systematic implementation of SoC security policies

2.2.1 IP Security Wrappers

Security wrappers extract security-critical events from the operating states of the underlying IP for communication with E-IIPS. Note that the naive approach of simply extracting all data, control, and status signals from IPs to E-IIPS would incur prohibitive communication and routing overhead. To address this problem, we develop security wrappers on IPs that incorporate "smartness" to detect security-critical events of interest in the IP while providing both a standard communication interface between the IP and the E-IIPS and a standard template-based design that can be easily integrated on top of the IP implementation.

How can the wrapper identify security-critical events while still providing a standardized template-based design? The key observation is that IPs in an SoC can be divided into a small collection of broad categories and security-critical events. Table 2.1 shows some of the broad IP categories together with some of the security-critical information relevant to each. For instance, "Memory IPs" include all IPs controlling the access to different memory hierarchies, e.g., memory controllers, Direct Memory Access (DMA) modules, cache controllers, flash control logic, etc., and processor cores include general purpose CPUs, GPUs, as well as cores controlled by microcode/firmware, e.g., Ethernet, UART controller, audio, video cards, etc. The security-critical events of interest also standardize substantially within each IP category. For instance, events in a Memory IP include read and write requests to specific address ranges by particular IPs (including DMA), as well as functional/standby modes. On the other hand, events in a processor core include start and end of critical system or application processes and threads, computations generating exceptions, interrupts by system controllers (often utilized by adversaries to jump to critical system addresses), etc. Finally, any event is associated with metadata that is sufficient for information about the event, e.g., DMA access details

Table 2.1 Representative set of security-critical events according to IP type

Type of IP	Example IPs	Type of events	Associated metadata
Memory IP	Memory/cache controller, DMA engine	Read/write request to specific address, DMA access, execution mode	Page size, burst size (DMA) ECC type, low power clock rates
Processor core	CPU, GPU, Ethernet controller audio and video card	Start/end of critical system threads, system interrupt, firmware upgrade request	Stored operation logs, flag or register settings, process duration
Communication core	Bus Controller, Bridge, Router, USB controller, PCI Express	Data transfer request, source/destination address peripheral IP transfer request, idle modes	Transfer packet size, serial frequency arbiter priority, bus clk. rate
Hard logic custom IP	AES, SHA engines, FFT, DWT block	Firmware integrity check start/end secure key access, FFT request by video card	Duration of operation, local clk domain

can be analyzed from page size, DMA burst size, and address range. This metadata is communicated by the security wrapper to E-IIPS, often under request from the E-IIPS (see below). Of course, in addition to standardized events, there are some IP-specific requirements in each IP. Our framework allows the SoC integrator to request additional security-critical events from specific IP which can then be mapped into its wrapper.

2.2.2 Security Wrapper Implementation

Our security wrapper design is frame-based, with a standard format for security-critical event definitions, which can be instantiated into corresponding events for specific IPs. A typical IP security wrapper architecture is shown in Fig. 2.2. Figure 2.3a illustrates an event frame. The wrapper typically consists of an activity monitor logic (to identify whether the IP is active), an event type detector, and a buffer to store the event metadata. Some events require also corresponding local clock domains. The wrapper also incorporates registers (see below) that would be configured by E-IIPS at boot time to specify particular events which

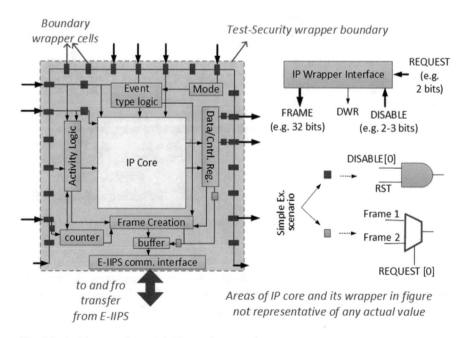

Fig. 2.2 Architecture of a generic IP security wrapper

A Frame of Event Log

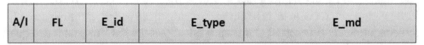

A/I	FL	E_id	E_type	E_md

Could be more than 1 frame collected per event depending on IP type, system designer and/or run time requirements

(a)

A/I – IP active or Inactive

FL – Frame Length

E_id – The IP event ID with which frame is linked

E_type – Type of IP event

E_md – Metadata about event

(b)

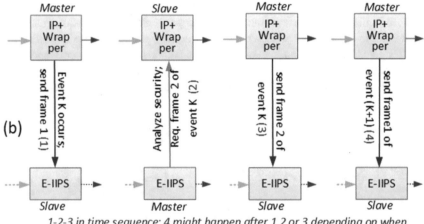

1-2-3 in time sequence; 4 might happen after 1 2 or 3 depending on when event (K+1) of interest occurs and specific design details of IP & E-IIPS

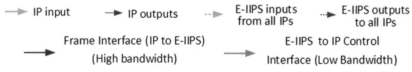

IP input   IP outputs   E-IIPS inputs   E-IIPS outputs
                                       from all IPs   to all IPs

Frame Interface (IP to E-IIPS)   E-IIPS to IP Control
(High bandwidth)   Interface (Low Bandwidth)

Fig. 2.3 (**a**) Fields of a typical event frame; (**b**) an example communication protocol between wrapper and security engine

need notification. The frame-based interface is used to provide a standardized communication mechanism with E-IIPS. In general, E-IIPS provides two types of signals to a security wrapper: (1) `disable` to block IP functions (in varying granularity depending on policy) in response to a suspected security compromise, and (2) `request` to request more data or send controls. We exploit the existing boundary scan interface of the IP to transmit data in parallel shift/access mode in high-bandwidth demands for certain functional security validation.

2.2.3 Security Policy Controller

E-IIPS acts as the "security brain" of the SoC, providing programmable interface to different security policies. Its key functionality is to analyze events communicated by the security wrappers, determine the security state of the system, and communicate IP-specific `request` and `disable` signals. Figure 2.4 shows the top-level architecture of E-IIPS. The architecture includes the two major components, viz., (1) a *Security Buffer* that provides access to the IP-specific event logs from the security wrappers, and (2) the *Policy Enforcer* that forms the analysis component of E-IIPS.

Security Buffer The security buffer interfaces with the Policy Enforcer through a *buffer controller* that defines how the buffer frames are analyzed by the Policy Engine. We implement the buffer storage through a standard static segmentation scheme, permitting variable-length segments based on the volume of metadata. The event logs can be read by the controller through ports on the buffer (controlled by the buffer controller). The IP-buffer control logic maintains synchronization and coherence of the security wrapper and Control Engine with data frames from IPs with different read and write speeds, segment sizes, and event frequency.

Policy Enforcer We implement the policy enforcer as a microcontroller-based implementation, which can be performed on a standard processor core. Functionally, the enforcer is a microcontrolled state machine that asserts or deasserts the required `disable` or `request` signals for different IPs. In addition to a microcontrol engine, it also includes a standard instruction memory (for storing microcode or firmware implementing the policies), and a small amount of data memory for intermediate computation. The nature of the computation involved in security policy

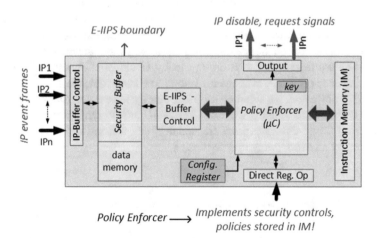

Fig. 2.4 Central security policy controller architecture

enforcement requires some custom modifications of existing commercial cores. We summarize a few of the illustrative necessary modifications.

- *Direct Register Writes.* Modifications to processor register file update logic is made to allow direct register updates, avoiding extra cycles for instruction and operand fetch from memory. This is necessary for time-sensitive policies, including TOCTOU and some access control policies.
- *Fused Data paths and Secure Mode.* Often the event metadata width of an IP permits fusing the data path (e.g., two 32-bit registers into one 64-bit) which facilitates concurrent analysis of multiple frames.
- *Branch Prediction Buffers.* Branch prediction buffer design is a critical require-ment for achieving low power and performance overhead, since security policy implementations involve the use of conditional branches with a much higher frequency than traditional application and system programs.

Finally, the E-IIPS module includes configuration register to permit the SoC designer to activate only a subset of implemented policies for a specific application or use case. The register is configured at design time through a combination of fuse/antifuses and multiplexers. It also aids in extending E-IIPS as a plug-n-play standalone IP with a generic architecture.

2.2.4   Secure Authenticated Policy Upgrades

Our threat model permits attacks through malicious firmware or software to subvert protection of system assets. Since E-IIPS itself is microcontrolled, it is also vulnerable to such attacks. Unfortunately, it is not possible to protect E-IIPS by merely disabling updates to its firmware. Since a key reason for a microcontrolled design is to permit policy upgrades on-field, it is critical to permit such upgrades through firmware updates. To address this problem, we implement an authentication mechanism based on challenge–response keys. Keys are generated at power-on using a standard technique based on Physically Unclonable Functions (PUF), that exploits intrinsic process variations on silicon to ensure robustness. Since keys are generated at power-on, we avoid on-chip key storage access control attacks through software or firmware. Finally, we avoid TOCTOU attacks during firmware updates by requiring single-threaded firmware updates, i.e., a firmware update cannot be interrupted by an overlapping update request.

2.2.5   Policy Implementation in SoC Integration

Security policy implementation through our framework requires collaboration between IP developer and SoC system integrator.

IP Provider The IP provider identifies the key standard security-critical events in the IP, based on the IP type (Table 2.1); this content is incorporated into frames along with registers for configuration, to build the security wrapper.

SoC Integrator The SoC integration team implements the security policies chosen for the application through the Policy Enforcer firmware of E-IIPS. Furthermore, since E-IIPS is centralized and IPs are delivered with security wrappers integrated, the SoC integration validation team is responsible for verification of policies against system-level use cases through simulation, hardware accelerator, FPGA, or post-silicon.

2.3 Use Case Scenarios

We present two use case scenarios of how generic security policies in the domain of secure-crypto and access control are mapped into our proposed architecture. The policy type, its function, and the involved IPs in our implementation are shown in Table 2.2.

2.3.1 Use Case I: Secure Crypto

Policy Crypto-processor data paths including encryption engine need to be functionally validated at power-on before any execution.

The above policy ensures trustworthiness of the system at boot time. The validation stipulated by the policy includes checks for correct operation of encryption (e.g., AES) and hash (e.g., SHA-1) engines as well as ensuring the stochasticity (randomness) of bits output by the True Random Number Generator (TRNG). Often, an undetected functional failure of the crypto data path, etc., results in compromise of the system security, mostly in terms of availability of resources (leading to denial-of-service attacks, etc.). In this study, we provide a sample implementation of how an SoC designer maps the AES engine verification into the proposed platform. The flow of operations/messages between the E-IIPS and IP security wrappers through the standard interfaces is illustrated in Fig. 2.5.

Table 2.2 Policies for usage case analysis

Sec. policy	Desc.	IPs involved
Secure crypto verification	Verify functionality of AES engine at power-on	E-IIPS, mem. cntrllr, crypto-proc., test-access cntrl.
Access control	Prevent DMA access to system-level addresses	E-IIPS, mem. cntrllr, DMA

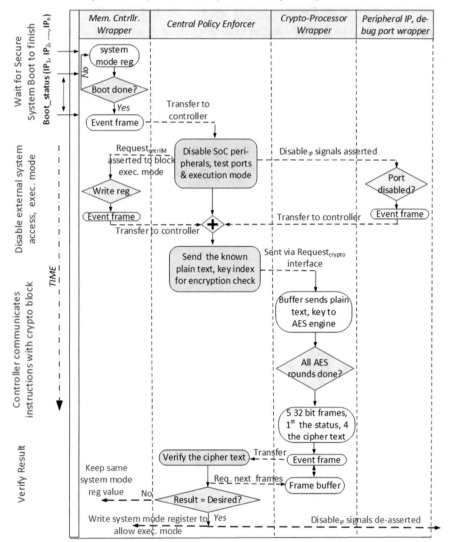

Fig. 2.5 Flow/message diagram representation of implementation of case I

In our implementation, the E-IIPS waits for the system boot process to finish (including power-on-self tests, firmware integrity check, and system software load to memory) before proceeding with the AES verification. This ensures the full trustworthiness of other system components during the check. The "boot mode finish" is indicated by a particular value of system mode register, mapped to the memory. In response, the E-IIPS disables external system interfaces of peripheral cores, JTAG, and other test/debug ports to eliminate possible attack surfaces. It also blocks transition to system execution mode. E-IIPS configures a set of known plaintext and key inputs in a buffer in the crypto-processor wrapper at boot time through potentially using the serial/parallel boundary scan interface for high-bandwidth communication (not shown in Fig. 2.5). The appropriate crypto test access port settings are asserted by the JTAG/TAM controller, in response to E-IIPS configuration request during boot. The desired cipher outputs are stored inside E-IIPS boundary. E-IIPS sends a particular plaintext/key buffer index to crypto-wrapper for execution. The computed cipher text is communicated through frames (or boundary scan) to the E-IIPS for verification. If it matches the desired, the proactive holds are lifted and the system goes to the normal execution phase. The exact sequence in terms of event detection and message communications is summarized:

- Memory Control Wrapper (MCW) detects event "system boot finish" and transfers frame to E-IIPS.
- E-IIPS reads the event from security buffer and asserts the disable interface to peripheral cores and test/debug ports. It blocks system execution by write to register mapped in memory through the MCW request interface.
- Receiving confirmation through frames that these actions have been performed by IP wrappers, the E-IIPS sends the plain text and key buffer index through the crypto-processor wrapper (CPW) request interface.
- The cipher text is computed, 5 frames are generated inside CPW, the first one containing the event "Encryption Complete" and metadata indicating that next 4 frames (of 32 bits) constitute the 128-bit cipher output.
- CPW sends the frame 1. E-IIPS sets the appropriate CPW request signals for the next 4 frames.
- E-IIPS verifies the computed cipher text.

This also shows a use case where the P1500 boundary scan infrastructure can be suitably used for high-bandwidth data/control communication in our framework, thereby reducing routing complexity and overhead.

2.3.2 Use Case II: Access Control

Policy Direct Memory Access (DMA) is prohibited in system-specific (ring 0/1 in 4-ring system) addresses of different IPs in the SoC memory space.

Most current SoCs involve DMA to the system memory through a dedicated DMA controller to reduce the workload on the processor cores. DMA by I/O

"DMA access is prohibited in system specific (ring 0/1 in 4 ring system) addresses of different IPs in the SoC memory space"

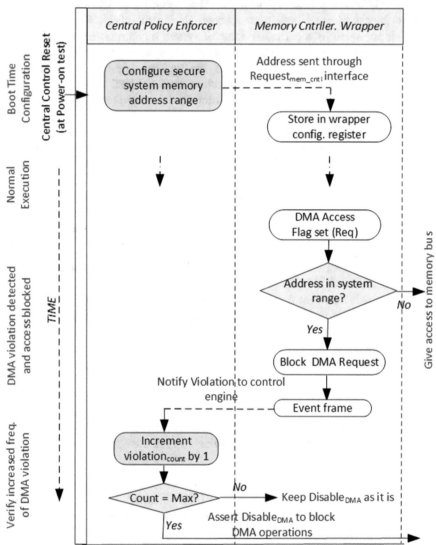

Fig. 2.6 Flow/message diagram representation of implementation of case II

peripherals (in memory-mapped I/O schemes) are utilized by attackers to snoop assets and modify system-level code. Policies like the one above protect against these security threats.

As illustrated in Fig. 2.6, E-IIPS configures the IP-specific system-level address ranges at boot time in the memory controller through its security wrapper (MCW).

When an access from the DMA controller is detected by MCW, the requested address is checked with the system-specific ranges inside the wrapper logic. In case of no violation, the system memory bus is granted for DMA. In case of system address overlap, the request is blocked, the violation is logged as an event, and is communicated to the E-IIPS through frames by MCW. E-IIPS maintains a buffer of DMA violations in recent past. If the number exceeds a threshold within a set time (configured by SoC designer), memory access requests from the DMA controller are disabled. The specific events/message flows and interface signals, as shown in Fig. 2.6, are summarized:

- E-IIPS configures system address ranges in memory controller register through MCW request interface at boot.
- When a DMA request is detected, the MCW checks the corresponding address. If violation is detected, the request is blocked. The event is sent as a frame to E-IIPS.
- E-IIPS updates count of DMA violations and compares with threshold. If the number exceeds within set time limit, the DMA controller operations are disabled through its disable interface.

2.4 Overhead Results

Given the dearth of appropriate open-source SoC design models to perform experiments, we are in the process of implementing a simple SoC design to assist in the current research. Our toy model has IPs of different functionalities interacting with each other to perform specific system functions. The IPs are obtained from opencores [49] in Verilog RTL models. The current version of this model at the time of this writing is illustrated in Fig. 2.7. In this model, all memory are implemented as register files for ease of synthesis. IP address spaces are mapped to the memory. These IPs can access the memory directly, analogous to DMA. At present, all IP–IP communication are point-to-point. The model has been functionally validated in ModelSim.

All IPs are wrapped with security abstraction layers. Events detected by these wrappers include a major representative subset of those listed in Table 2.1, e.g., read/write requests to memory, duration of processes, specific conditional jumps in μP core, transfer start/end of SPI module, etc. The 32-bit frame formation logic and metadata buffers are present. IPs contain configuration registers which are configured by E-IIPS at boot time (master system reset asserted). The E-IIPS is implemented with a single DLX 32-bit RISC core (5 stage pipeline). Firmware (security policies) is stored in local instruction memory of 4 KB. 2 bit disable and request signals are output to each IP from the controller.

To obtain representative overhead values, the IPs were synthesized at 32-nm predictive technology library. The calculated area, dynamic (at 1 GHz clock), and leakage power overheads are provided in Table 2.3. The overheads are mostly

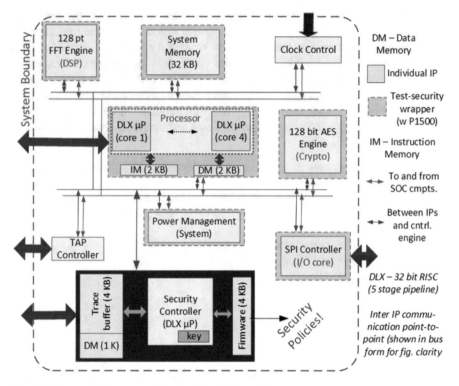

Fig. 2.7 Current architecture of functional toy SoC model

minimal. In some scenarios, the power reduces after resynthesizing with wrappers due to internal (heuristic) optimizations and hence reported as negligible. E-IIPS was synthesized at 32 nm and the resulting area and power (at 1 GHz clock) values are provided in Table 2.4. Finally, the area overhead of the control engine was estimated with respect to our toy model and commercial SoCs from Apple and Intel at 32 nm (~1 GHz clock) and provided in Table 2.5. The 32-KB system memory (major area component) area was estimated with established SRAM models. As our toy SoC is rather small with only handful of IPs, the overhead is comparatively higher. The overhead for the controller is minimal in realistic scenario. For a generic SoC design, we can conclude that the hardware overheads due to the proposed architecture would be minimal. After analysis with NoC fabrics of different types, the routing complexity and transfer power/energy would be evaluated as part of the future work.

Table 2.3 Area and power overhead of IP security wrapper (at 32 nm)

IP	Orig. area (μm^2)	Area ovhd (%)	Dyn. Pw. ovhd (%)	Leak. Pw. ovhd (%)
AES (128 bit)	101,620	2.1	–	–
SPI controller	3947	9.2	11	9.7
DLX μP core (w. 4 KB I/D mem.)	290,496	6.8	–	5.4
FFT (128 point)	1810	10.2	–	16.1

–: negligible

Table 2.4 Area and power of central security controller (at 32 nm)

Die area (μm^2)	Dynamic power (mW)	Leakage power (mW)
2,831,860	13.67	34.13

Table 2.5 Die area overhead of central controller (at 32 nm)

SoC	Die area (μm^2)	Overhead of controller (%)
Our toy model	13.1×10^6	21.7
Apple A5 (APL2498)	69.6×10^6	4.06
Intel atom Z2520	$\sim 40 \times 10^6$	7.1

2.5 Conclusion

We have presented a novel architectural framework for implementing diverse security policies in a system-on-chip. It enables systematic implementation of security policies, and hence can greatly facilitate the process of secure SoC design involving various security assets and can potentially reduce the design/hardware overhead. Furthermore, it enables effective validation and debug and update of security policies during post-silicon validation, which often imposes major roadblocks in SoC production cycle. The architecture consisting of a centralized security policy controller, referred to as E-IIPS, and generic security wrapper per IP block, is easily scalable to a large number of IPs and flexible to accommodate IP blocks of varying structural properties. The proposed E-IIPS module builds on extends the functionality of recently reported infrastructure IP for security, which was proposed for implementing multitude of security functions (e.g., protection against scan-based attack, Trojan attacks, power analysis attack, and hardware piracy) using a shared centralized architectural fabric. The E-IIPS module we have presented in this paper, when combined with the capability of implementing conventional hardware security functions, can serve as even more powerful security infrastructure IP. We have verified functional correctness of the architecture through extensive simulations and evaluated the hardware overhead. The hardware overhead for the proposed architecture is expected to go down significantly for realistic SoCs.

2.6   Bibliographic Notes

The notions of high-level security requirements in a computing system were developed in the 1990s as part of research on information security [26]. Early research on security policies looked primarily on software systems and developed analysis frameworks for access control and information flow policies [24, 30]. More recently, researchers have tried to develop languages for formal representation of hardware security policies [42]. On the other hand, with the increasing prominence of SoC designs, there has been significant research interest in SoC security. However, most recent research in this area has focused on *hardware security*, i.e., protection of the system against a malicious hardware Trojan [13], various forms of counterfeiting attacks [27], and attacks to leak secret information through side channels or on-chip resources such as scan or debug infrastructure. A recent work has also targeted developing a centralized and scalable framework, referred to as an infrastructure IP for security, for efficiently realizing countermeasures against these attacks [73]. Infrastructure IPs refer to a range of IPs that are dedicated to facilitate SoC functional verification, testing, or yield improvement [77]. On the contrary, the proposed solution can be viewed as an extended infrastructure to implement system-level security policies in SoC. There are also a number of works that report efficient protocols [39] involving functional IP blocks, crypto-IPs, and security wrapper and associated architecture-level custom modification for specific security policies [64]. However, they do not propose a generic architecture involving a reusable centralized IP and its flexible interface with the IP blocks, as proposed in this paper.

A preliminary version of this work appeared in the International Conference on Computer-Aided Design in 2015 [6].

Chapter 3
Exploiting Design-for-Debug in SoC Security Policy Architecture

3.1 Introduction

In the last chapter, we observed that systematic, methodical implementation of SoC security policies typically involves smart wrappers extracting local security-critical events of interest from IP blocks, together with a central, flexible control engine that communicates with the wrappers to analyze the events for policy adherence. It was analyzed that for certain complex, security-critical constituent IP modules, considerable hardware overhead may be incurred in designing these wrappers according to the policy requirements. Although typically, wrappers follow a standard template-based design based on the type of IP, some custom modifications would be required to adapt it to the particular SoC usage scenario and the role of the IP in the SoC architecture. Along with resource overhead, these may also lead to increase in design complexity and associated time-to-market. In this chapter, we consider how to solve this problem.

Our approach is to exploit the extensive design-for-debug (DfD) instrumentation already available on-chip. Modern SoC designs contain a significant amount of DfD features to enable observability and control of the design execution during post-silicon debug and validation, and provide means to "patch" the design in response to errors or vulnerabilities found on-field. Hence, typically the DfD modules already detect the information necessary for most of these events required for security policies. On repurposing the debug infrastructure for security, the DfD trace macrocells local to the IPs can be configured to extract these security-critical events during SoC normal execution. In addition to reduction in the overall hardware overhead, the proposed approach also adds flexibility to the security architecture itself, e.g., permitting use of on-field DfD instrumentation, survivability, and control hooks to patch security policy implementation in response to bugs and attacks found during post-silicon or changing security requirements on-field. In this chapter, we demonstrate how to design scalable interface between security and debug architectures that provides the benefits of flexibility to security policy

© Springer International Publishing AG, part of Springer Nature 2019

S. Ray et al., *Security Policy in System-on-Chip Designs*,

https://doi.org/10.1007/978-3-319-93464-8_3

implementation without interfering with existing debug and survivability use cases and at minimal additional cost in hardware resource, energy, and design complexity. Below, we provide a brief background on typical debug infrastructures implemented in modern SoCs.

3.2 On-Chip Debug Infrastructure

The supported functionality, integration density, and complexity of modern day System-on-Chips (SoCs) have increased manifold over the years. At the same time, the number of different, heterogeneous hardware–software-based IP blocks have grown inside an SoC. Together with extremely aggressive time-to-market schedules in the SoC ecosystem involving various stakeholders, these factors have made post-silicon validation and debug of SoC designs the most complex, tedious, and difficult part of the design process. Design-for-Debug (DfD) refers to on-chip hardware for facilitating post-silicon validation [69]. A key requirement for post-silicon validation is observability and controllability of internal signals during silicon execution. DfD in modern SoC designs include facilities to trace critical hardware signals, dump contents of registers and memory arrays, patch microcode and firmware, create user-defined triggers and interrupts, etc. As an estimate, on-chip debug infrastructure typically comprises ~20–30% of the total silicon die area of a modern day SoC [69]. Furthermore, the DfD architecture is getting standardized to enable third-party EDA vendors to create software APIs for accessing and controlling the hardware instrumentation through the debug access ports, for system-level debug use cases. As an example, ARM CoreSight^{TM} architecture [14] provides facilities for tracing, synchronization, and time-stamping hardware and software events, a trigger logic, and facilities for standardized DfD access and trace transport.

A representative block diagram schematic of a debug infrastructure in an SoC, along the lines of ARM CoreSight^{TM} is illustrated in Fig. 3.1. As shown, typically it comprises of the following major components—(1) Access Port; (2) Local Trace Sources with trigger logic; (3) Trace Sinks/Hub; and (4) Debug Communication Fabric. The access port provides access to an external debugger or an on-chip memory debugger to configure the individual local debug component triggers for hardware/software tracing. External connection could be through a standard JTAG or Serial Wire connection. Internally, the access to trace source configuration registers (programmer visible) is memory mapped through, for example, a debug configuration bus, controlled at the access port. JTAG-based scan, if present, is also controlled from the corresponding access port. The local debug logic, sprinkled around the SoC and instrumenting individual IP activity/events/traces, comprise the trace sources, e.g., Embedded Trace Macrocell (ETM), System Trace Macrocell (STM) in CoreSight. They comprise hardware or micro-controlled logic like address, data comparators, performance counters, event sequencers, logical event combinations, etc., that can be used to detect a particular configured event and

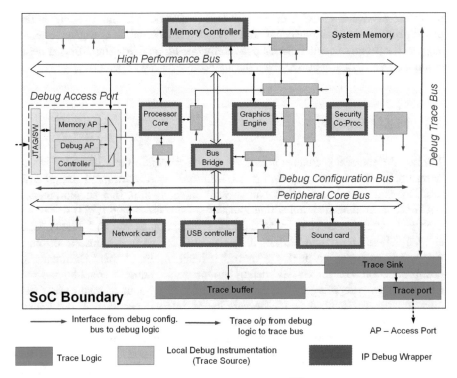

Fig. 3.1 Simplified SoC DfD architecture based on CoreSight^{TM}

thereby trigger collection of hardware/software traces around the event (analogous
to breakpoints, and watchpoints in simulation). A simple example for a μP core
could be detection of the program counter traversing specific system memory
address ranges and tracing all executions around it. In this scenario, the event
and the corresponding start and end address ranges would be configured by the
debugger accordingly. The trigger for tracing might be local to the IP block or
global, arising from other IP debug sources (cross-trigger). Finally, the traces are
transported over the trace bus according to standard protocols (e.g., MIPI-STP [20])
and communicated externally through the trace port or stored in trace buffers. The
trace sink logic controls the trace collection with source IP encoding and time
stamping for synchronization.

To interface with the local debug instrumentation, IP blocks are augmented with
standard debug and test wrappers (by IP providers) which extract out the critical
nets, IP traces, registers, and other important design features, while abstracting out
the internal implementation details of the IP. Besides providing observability into
the design, on-chip debug often also provides controllability to the IP functionality
to enable patches and upgrades on field and thereby supports *survivability*. Typically, the full range of on-chip debug features is not required after design passing
full post-silicon validation. But, they are still mostly kept in the final production

ready designs due to probable changes in critical path delays, routing complexities, power profiles, etc., that may occur after any design modification as well as to ensure support for debug of critical issues potentially arising on field. The different debug components are usually power gated during normal on-field operations.

3.3 Methodology

The key insight of this chapter is that we can implement a security policy control framework without incurring significant additional architecture and design overhead, by exploiting infrastructures already available on-chip. In particular, modern SoC designs contain a significant amount of Design-for-Debug (DfD) features to enable observability and control of the design execution during post-silicon debug and validation, and provide means to "patch" the design in response to errors or vulnerabilities found on-field. On the other hand, usage of this instrumentation post-production, i.e., for on-field debug and error mitigation, is sporadic and rare. Consequently, computing systems have a significant amount of mature hardware infrastructure for control and observability of internal events that is typically available and unused during normal system usages.

The main contribution of the chapter is a flexible security architecture that exploits on-chip DfD features for implementing SoC security policies in a systematic fashion. This refines the architecture framework of the previous chapter with the interface between the security and on-chip debug infrastructures to make it more light weight and flexible. According to the requirements of the security policy architecture, IP security wrappers must be *smart*, i.e., they would detect a standard set of security-critical events of interest depending on IP type as well as some custom ones based on the particular SoC usage scenario. The events necessary are based on the policy as well as the IP involved, e.g., for a CPU typically, we need to detect attempts of privilege escalation and monitor control flow of programs to detect the probable presence of malware in different stacks as well as prevent fine-grained timing-based masquerading attacks. Developing IP security wrappers thus requires custom hardware logic for identification of these events. Sometimes in the scenario of a large, complex, security-critical IP, the resource overhead of the wrappers may be considerable. However, the DfD modules on chip already detect the information necessary for most of these events. For example, Table 3.1 illustrates a few representative security-critical events that can be detected through CoreSight[TM] macrocell for a processor core. Here, the macrocell is assumed to implement standard instruction and data value/address/range comparators, condition code/status flags match check, performance counters, event sequencers, logical event combinations, etc. Correspondingly, local DfD for NoC fabric routers can detect bulk of the critical events required for addressing threats such as malicious packet redirection, IP masquerade, etc. We show how to build efficient, low-overhead security wrappers by repurposing the debug infrastructure, while being

Table 3.1 Typical security-critical events detected by DfD trace cell in processor core

Trigger event	Ex. security context
Program counter at specific address, page, address range	Prevent malicious programs trying to gain elevated privileges
System mode traps for specific interrupts, I/O, file handling, return from interrupt	Verify limited special register access by other IPs in kernel mode
High conditional branch or jump instruction frequency	Highly branched code often signs of malware
Invalid instruction opcode, frequent division by exceptions	Untrusted program source; apply strict 0 access control
Read/Write request to specific data memory page/s	Protect confidentiality, integrity of security asset
# of clock cycles between 2 events = threshold	Satisfy resource availability and avoid deadlock
More than one inter communicating threads	Verify TOCTOU policy in authenticated firmware load in μC

transparent to debug and validation usages. We illustrate some of the design trade-offs involved between complexity, transparency needs, and energy efficiency.

3.4   DfD-Based Security Architecture

Our architecture is built on top of the E-IIPS framework developed in the previous work [6]. In particular, we exploit DfD to implement smart security wrappers for IPs that communicate with the centralized policy controller (SPC or E-IIPS), which implements security policies. Furthermore, SPC can also program DfD to implement security controls using defeature and control logic available for on-field patching and upgrades.

3.4.1   Debug-Aware IP Security Wrapper

We architect smart security wrappers for each IP by exploiting DfD to identify relevant security-critical events; on detection of such an event, DfD communicates the information to the IP security wrapper that communicates it to the centralized SPC. Direct communication between DfD and E-IIPS requires the trace communication fabric to be active along with appropriate interface between the trace sinks and E-IIPS block. As heavy weight trace fabrics are usually power gated during normal

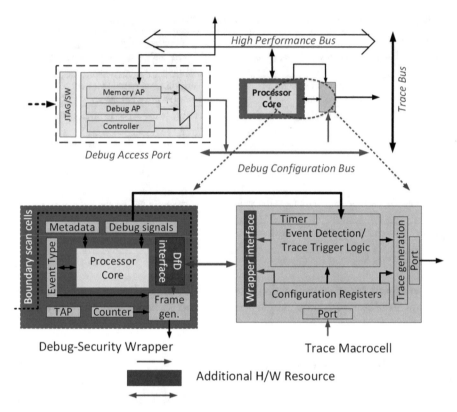

Fig. 3.2 Additional hardware resources for interfacing DfD with IP security wrapper

system operation, this direct DfD–SPC communication may lead to significant power overhead and additional hardware requirements as well. Hence, the DfD detected events would be sent to E-IIPS through the corresponding IP security wrapper. To enable this functionality without hampering debug usages for the DfD, we need local (IP-level) modification of DfD logic and appropriate adaptation of the security wrapper. Figure 3.2 illustrates a block diagram level schematic of the additional hardware requirements. In particular, noninterference with debug usage requires transmission of security data to SPC via a separate port (instead of repurposing the debug trace port), which requires an additional trigger logic.

The events of interest for the IP are programmed by the SPC via the configuration register interface of the corresponding DfD module. Since DfD module can be configured to detect a number of security events (related or disparate) at runtime, SPC must correctly identify the corresponding event from the communication frame sent by the security wrapper. We standardize this interface across all local DfD/security-wrapper pairs, by tagging event information with the corresponding configuration register address. We note that this standardization comes at the cost of additional register overhead in IPs where one or only a few events are detected via debug logic. Trace packet generation controls can be disabled during SPC access (similarly, security-debug interface disabled during system debug) to save leakage power when debug (similarly, security) architecture is not in use. Besides security-critical event triggers and observability of associated information, the local DfD control hooks can also be repurposed (by appropriate SPC configuration) to enforce security controls in the IP via the existing debug wrapper, during both design and on-field patch/upgrade phase.

3.4.2 SPC-Debug Infrastructure Interface

The Security Policy Controller (SPC) must be able to configure the individual local (to the IPs) on-chip debug logic to detect relevant security-critical events and assert appropriate controls. Figure 3.3 illustrates the communication between SPC and the debug interface along with additional modifications required for this interface. As used during system debug, the existing configuration bus (address and data) is used by SPC for trace cell programming. For enabling SPC access to the configuration bus, small enhancements are necessary in the debug access port (DAP) to include an SPC access link and potentially scaling up the control logic (shown as simple multiplexer here in representation) for DfD configuration source selection. This incurs minimal hardware overhead in comparison to typical debug infrastructure resources found in SoCs.

As there are usually enough configuration registers and associated logic in the local DfD components to monitor all possible security-critical events, the SPC can configure the trace cells with appropriate values at boot phase; therefore, the configuration fabric can be turned off during most times to save leakage power. In some rare scenarios, SPC cannot configure DfD for detection of all necessary events at boot; for these cases, SPC interfaces with the power management module to turn

Fig. 3.3 Interfacing SPC with on-chip debug

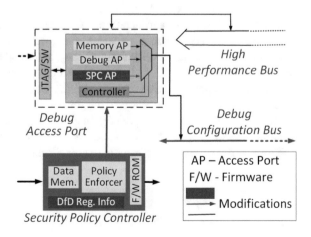

on the Debug Access Port (DAP) and configuration bus at runtime. Apart from the incorporation of the instructions in the SPC firmware memory to load different configuration register values at boot and potentially run time, the programmed register addresses (or other identifier) and associated data values are stored over run time in the SPC data memory (shown as SPC modifications in Fig. 3.3). These are used to uniquely identify DfD detected events from frames sent through the corresponding IP security wrappers over system execution. As an example, if the event frame consists of the configuration register address associated with DfD detected event, the SPC can match/verify with the stored database and uniquely identify the event.

3.4.3   Design Methodology

An SoC design flow involves integration of a collection of (often predesigned) IPs through a composition of NoC fabrics. An architectural modification involving communication among different IPs typically disrupts the SoC integration flow. We now outline the changes necessary to SoC integration to adapt to our proposed DfD-security architecture.

IP Designer   The IP provider needs to map the required security-critical events to the DfD instrumentation for the IP. The respective configuration register values are derived from the debug programming model. Finally, the security wrapper is augmented with custom logic for events not detected by DfD, and the standardized event frames and interface for communicating with SPC are created, along with the wrapper to DfD interface.

SoC Integrator   The SoC integrator augments the event detection logic of the local DfD instrumentation in IPs with appropriate triggers to the DfD/security-wrapper interface for transmission of event occurrence information, modifies debug access port with required hardware resources to incorporate SPC access to debug, and adds necessary security and debug access control requirements to ensure debug transparency in the presence of security requirements. For the latter use case scenario, i.e., ensuring security during debug and validation, where the DfD may not be repurposed to detect all security-critical events, the SoC integrator may proceed with more proactive, stricter security controls for constituent IPs in the SoC operations. All the necessary configuration register addresses/values are stored in the additional data memory/buffer in SPC to uniquely identify DfD detected security-critical events. The SPC is also augmented with firmware instructions to configure these debug registers, mostly during boot.

3.5 Use Case Analysis

3.5.1 An Illustrative Policy Implementation

To illustrate the use of our framework, we consider its use in implementing the following illustrative policy:

I/O Noninterference When the CPU is executing in high-security mode, I/O devices cannot access protected data memory.

The policy, albeit hypothetical, is illustrative of the typical security requirements involved in protecting assets from malicious I/O device drivers. Figure 3.4 illustrates the flow of events involved in the implementation of this policy by SPC through DfD/security wrapper of the associated processor. Here, the DfD configuration is through a debug access port, CPU has an associated Embedded Trace Macrocell as the local DfD instrumentation, and I/O device requests are assumed to be based on Direct Memory Access (DMA). Following are the key steps involved:

1. During boot, SPC configures the required DfD instruments. This includes "*program counter within secure code range*" and "*write access to protected data memory*" event triggers in the ETM.
2. Along with DfD, the SPC configures the IP security wrappers for the subset of events to detect, frame protocol to follow, scan chain access, etc. DMA engines are also configured by boot firmware on device–channel and channel to memory mapping.
3. When a secure program (assumed in protected code range) is loaded, the ETM detects the event, and triggers the security wrapper which communicates with SPC. The SPC updates the security state.
4. A DMA interrupt is detected by the corresponding security wrapper and transmitted to SPC. Any write request from the high-privilege driver to the protected memory is detected by ETM and transmitted to SPC via wrapper. The SPC identifies policy violation in context of the current security state and enforces necessary controls.

3.5.2 On-Field Policy Implementation/Patch

Given changing security requirements, e.g., adoption of system in different market segments or to address bugs or attacks detected on-field, new policies may require to be implemented or existing ones need to be upgraded or patched. These may require new events to be detected (outside what had been considered in design phase), extraction of more event information, and/or control the IP functionality in response to a policy. The interface of the security architecture with the on-chip debug infrastructure allows the possibility of on-field system reconfiguration and upgrade—something virtually impossible to perform with security wrappers

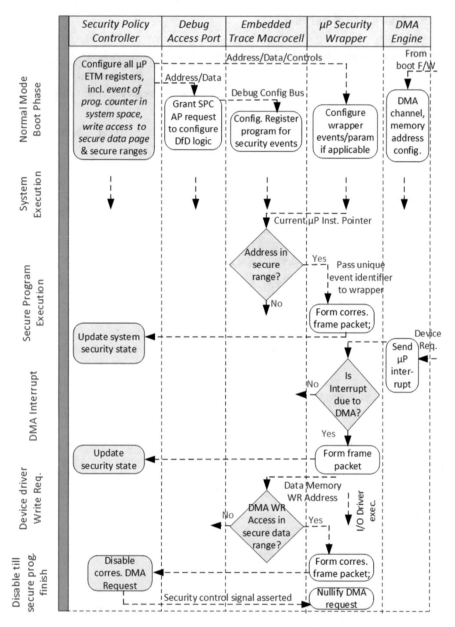

Fig. 3.4 Use case scenario of security policy implementation exploiting the local DfD instrumentation

customized for specific security policies. Achieving this requires selection of local DfD instrumentation at corresponding IPs to identify if the relevant events can be detected; if so, the corresponding register address and value are added to SPC memory to be sent through DAP at boot/execution for configuration. With a standardized DfD interface with the IP security and debug wrappers, the corresponding events can be uniquely detected and control signals asserted in the IP if applicable.

3.6 Experimental Results

As mentioned in the previous chapter, due to lack of standard open-source models for studying SoC architecture, we have been developing our own SoC design model. Although simpler than an industry design, our model is substantial and can be used for implementing realistic security requirements. Figure 3.5 shows the updated model (includes the representative debug infrastructure with interface to the security framework) from the version of the last chapter. It includes a DLX microprocessor core (DLX) with code memory, a 128b FFT engine (FFT), a 128b AES crypto core (AES), and an SPI controller. The IPs are augmented with security wrappers according to standard security-critical event set [6]. The SPC incorporates the DLX microprocessor core as the execution engine with policies stored in its instruction memory.

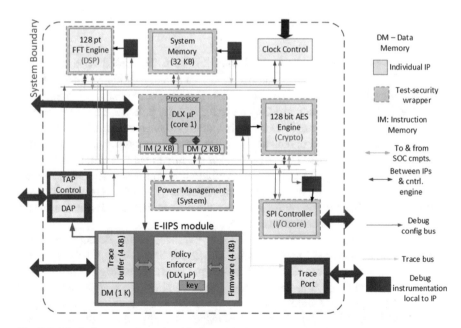

Fig. 3.5 Block diagram schematic of SoC model with on-chip debug infrastructure

Table 3.2 Example DfD instrumentation features by IP type in SoC model

DfD by IP type	Ex. DfD inputs	Ex. trigger events	Ex. trace content
DLX TM	Prog. counter (PC), Inst. opcode, Data RD/WR addr. (DA), Special condition codes	register value, PC in desired range, page, addr., Jump, Branch T/NT, Particular exc-eption, interrupt, DA in specific page	Past, future "N" inst. addr./values, status reg. values, next branch inst.
AES TM	Plain text (PT), Key, Cipher o/p (CT), Mode, Status, Intermediate round key	Encrypt/Dec. start/stop, Specific round reached, Key = desired	current 16B PT, Key, All future round keys, CT
SPI TM	Parameters, Status, i/p packet, Cycle counter, Configuration register	start/stop of operation, Source IP = desired, acknowledgement error	configuration register, past "N" i/p packets
Memory cont-roller TM	Addr., Data, RD/WR, Burst size, Word/byte granularity, ECC	Addr in specific range, bank, DMA request, change in row buffer	In/Out Data word/byte, future "N" data addr.

Table 3.3 Area (μm^2), Power (μW) of DAP (*SoC area* $\sim 1.42 \times 10^6 \mu m^2$; *SoC power* > 30 mW)

DAP area (orig.)	New DAP area	DAP Pwr (orig.)	New DAP Pwr
380.2	527.67	12.63	19.82

We implemented a representative debug infrastructure, based on a simplified version of ARM CoreSight™ features. It is functionally validated using Model-Sim [47] for typical use cases. Necessary interfaces/logic, as described in Sect. 1.6, are added to the model to support DfD reuse for security policies. The DAP controls memory-mapped accesses to DfD instrumentation via the configuration bus. It also contains logic to control simultaneous debug and security requirements. Local DfD, similar to CoreSight ETM/ITM, are added for the functional IPs with their features enlisted in Table 3.2. Each has 16 32-bit configuration registers (64B address space) and support interfaces with the corresponding IP security wrapper. On security event detection, the configuration register-based unique identifier (10 bits) is sent to the wrapper to be communicated with the SPC.

Table 3.3 summarizes the area and power overhead of the debug access port (DAP) to incorporate SPC access (which entails modification of DAP). The area estimation is provided from synthesis at 32-nm predictive technology. Note that with respect to base DAP design, the area and power numbers are high because of the simple DAP logic in the model; however, the additional system overhead induced by the modification is negligible since the DAP contribution to overall SoC die area

Table 3.4 Area (μm^2), power (μW) overhead of DfD trace macrocells in SoC

DfD TM	Die area (orig.)	Area ovrhd. (%)	Power (orig.)	Power ovrhd. (%)
DLX TM	15,617	6.07	512	6.7
AES TM	5918	8.5	165	10.9
FFT TM	2070	18.8	60.6	20.2
SPI TM	2054.6	17.08	57.75	15.8
Mem. TM	4623	7.9	163.3	1.65

Table 3.5 Area (μm^2) savings of IP security wrapper

Wrapper (corres. IP)	Orig. area	New area	Area savings (%)
DLX μP	3437	2326	32.32
SPI cntrl.	1055	842	20.2
AES crypto	1661	702	57.7

Table 3.6 Power (mW) analysis in SoC on implementation of debug reuse

IP	IP power consumption	Wrapper savings	Corres. TM power
DLX μP	6.54	0.52	0.551
SPI cntrl.	0.321	0.024	0.062
AES crypto	5.53	0.03	0.173

and power is minimal. The area and power overheads of DfD Trace Macrocells (TM) with respect to original TMs (without security wrapper interface) are enumerated in Table 3.4. The overheads are typically within 10%, but can be higher for some small IPs (e.g., FFT and SPI).

Table 3.5 measures the decrease in wrapper hardware area overhead through repurposing DfD for security wrappers. The measurement is done by comparing the current implementation with the earlier reference implementation in which the security wrapper is responsible for detecting all necessary security-critical events, with no dependence on DfD. We note that the savings can be substantial, ranging from 20% to close to 60%. This is because a comprehensive DfD framework typically captures a majority of the security-critical events since they are also likely to be relevant for functional validation and debug.

Finally, Table 3.6 measures power overhead. This experiment is interesting since DfD reuse has two opposing effects: power consumption may increase since trace macrocells remain active to collect security-critical events even when debug is not active; on the other hand, decreasing hardware overhead of the wrappers contributes to reduced power consumption. For DLX, the two directions cancel out, while for SPI and AES there is a net power overhead. We note that the overhead is minimal with respect to the overall power consumption of the entire SoC.

We end this section with an observation on interpretation of results. We note that the numbers provided are based on the security and DfD infrastructures in our SoC model; while reflective and inspired by industrial systems and policies, our implementations are much simpler. Nevertheless, we believe that the *overhead* measurements substantially carry over to realistic usage scenarios. Perhaps more importantly, our experiments provide the guidance on parameters to analyze when considering repurposing DfD for security implementation vis-a-vis standalone security wrappers.

3.7   A Potential for Hardware Patch

With the ever-rising functional capabilities and associated complexities of a modern day SoC, the number of different types of aforementioned security policies governing the access to on-chip security-critical assets is also increasing. With better standardized heterogeneous integration methodologies, optimized power management techniques on-chip along with the still ever-present shrinking transistor sizes, this trend of enhanced SoC resources and design complexity is expected to continue in at least the near future. Policy implementation typically involves subtle interactions between hardware logic, software, and/or firmware of the underlying IP blocks and other SoC components. Many of them also get refined or modified along the design cycle by various stakeholders. Besides, as compared to a significant fraction of these policies implemented at the operation system or system software level in typical processor-based computing platforms, majority of the SoC-level security policies involve direct interaction with the hardware logic of the constituent IP blocks. Obviously, a key challenge arises in the upgrade or patch of these policies when in-field. In this section, we consider preliminary observations regarding patching and how the E-IIPS framework can be used for that. We will explore patching more fully in Chap. 5.

Why do we need patching? This can be in response to design bugs found during post-silicon validation, for satisfying system-level performance/power profiles, or due to changing security requirements on field. The latter could arise due to vulnerabilities detected on-field or adoption of the final SoC-based system or product in different global market segments. Hence, in SoC designs, a key requirement during the upgrade is that of a patch in the underlying hardware logic rather than just a software patch/upgrade. The hardware patch may involve addition of new security policies or modification of existing system-level policies in the SoC. These could indirectly require extraction of more security-critical events from various IP blocks at a finer time–space granularity and/or setting of additional proactive controls than those considered at design time. On the other hand, it could also signify appropriate configuration to detect less events and/or apply subset of controls than originally designed with, to relax the security constraints.

The proposed centralized architecture framework (incorporating the interface to debug instrumentation), providing a systematic, methodical approach for security policy implementation in SoCs, is flexible and adaptable to be applicable towards the requirement of this hardware patch. The DfD trace macrocell, local to the IP block of interest, can be configured accordingly by the centralized security policy controller (SPC) to extract additional (to what was considered during design of security wrappers) internal security-critical events and/or set/disable appropriate controls to govern its functionality, depending on the system state. As mentioned in this chapter, debug infrastructure in a modern day SoC consists of a significant amount of resources for adequate observability/controllability, and thereby ensuring full coverage during post-silicon and/or on-field. Hence, the DfD blocks are capable of detecting majority or all of the security-critical events as well as controlling part of the IP functionalities, required for policy implementation. The necessary configuration register values can be found from the debug infrastructure programming model and stored in the SPC to be loaded at boot time. Without the DfD interface, one would not be able to implement these modified security policies (requiring extra hardware event logic/controls) without an additional fabrication cycle. Presently, in majority of these cases where a respin is not a feasible option, often the course of action is to implement pessimistic (safety first), proactive security policies which may hamper performance/power and vice versa depending on what is deemed critical. Within the existing hardware resources of the security wrapper of an IP, the event detection and trigger logic, metadata extraction, control logic modification, and frame generation parameters may be configured by the SPC (at boot) to adapt to changing (stricter or more relaxed) security requirements. This provision of the framework aids in hardware-level system security patch as well. Finally, the centralized SPC module, which at the high level is basically a state machine outputting different security controls after analysis of the system state, can be easily upgraded to implement new or modified policies utilizing these altered/additional extracted events.

A typical approach for designing SPC so that it is reconfigurable is a micro-controller based implementation of the SPC where the security policies are stored as firmware in a nonvolatile instruction memory. They would be upgraded based on secure authentication via on-chip keys. However, an alternative approach would be using a Field-Programmable Gate Array (FPGA) [41, 61] to implement the SPC module. An FPGA, which constitutes reconfigurable logic and interconnect fabric would also meet the SPC requirement of the upgradeability feature. The modified configuration bitstream would be uploaded to the internal FPGA nonvolatile memory during upgrade of security policies. Besides, for application scenarios like this centralized SPC module, which would not typically require frequent program–reprogramming iterations, an FPGA might be better suited especially with regard to implementation of time-sensitive policies like time-of-check-time-of-use (TOCTOU), liveness policies, etc., due to normally higher speed of operation of an FPGA as compared to a micro-controller. Besides as the underlying constituent hardware logic and their interconnections are changed during reconfiguration of an FPGA, they are more secure from design reverse-engineering-based attacks, which could be used by an adversary to gain knowledge of the policies (proprietary to the SoC design house) if possible. This change in constituent hardware logic of an FPGA may also aid in hardware-level patch during upgrade of security policies in an SoC. We explore this possibility more fully later in this book.

3.8 Conclusion

In this chapter, we have developed an SoC security architecture that exploits on-chip DfD to implement security policies. It provides the advantage of flexibility and on-field update of security requirements, while being transparent to debug use cases. This flexibility of hardware patch helps in scenarios of resolving bugs/vulnerabilities found on-field or changing security requirements, etc. Our experiments suggest that the approach can provide significant benefit in hardware and area savings of the IP security wrappers with no substantial energy overhead.

3.9 Bibliographic Notes

Early research on security policies looked primarily on software systems [24]. More recently, with the increasing prominence of SoC designs, there has been significant interest in SoC security. There have been research on exploiting DfD for protection against software attacks, e.g., Backer et al. [4] analyze the use of enhanced DfD infrastructure to confirm adherence of software execution to trusted model. Besides, Lee et al. [40] study low-bandwidth communication of external hardware with the processor via the core debug interface, to monitor information flow. Methods have also been proposed on securing SoC during debug/test by appropriate changes in DfD/test infrastructure [1, 3]. But, this work is regarding flexible and light-weight architecture framework for generic security policy implementation exploiting the repurposing of DfD for security-critical event extraction. Hence, it is complimentary to existing works in the domain.

A preliminary version of this research appeared in the proceedings of Design Automation Conference in 2016 [7].

Chapter 4
Security Assurance in SoC in the Presence of Untrusted Components

4.1 Introduction

The E-IIPS architecture discussed in Chap. 2 implements security policies to protect against attacks via rogue software stacks as well as threats from SoC to system external interface. However, the architecture did not account for malicious IPs (rogue hardware or firmware coming in from IP vendor). In this chapter, we exploit and extend the centralized policy implementation architecture in a systematic, disciplined manner to provide system-level security assurance in the presence of untrusted IPs. In particular, the proposed framework implements fine-grained IP-Trust aware security policies for run-time protection and mitigation of system-level vulnerabilities in an SoC in the presence of malicious logic both in the core IP functionality and corresponding security wrapper implementations.

4.2 Problem of Untrustworthy IPs

The increasing complexity of SoC design and validation coupled with strict time-to-market demands have typically led to SoC designers utilizing pre-qualified third-party IP blocks to increase design productivity [16, 43, 53, 76], as illustrated in a generic flow diagram in Fig. 4.1. These IPs could be of different types like processor core, graphics core, memory subsystem and corresponding controllers, device controllers as well communication fabric components. Test and debug frameworks are also integrated as infrastructure IPs. These IPs constitute one of the largest vulnerabilities in the present SoC design cycle. In particular, a modern SoC design typically constitutes several hundreds of IPs, most of which are procured by the SoC integration house from third-party vendors. With these third-party vendors located in different parts of the world with varying control over rules and regulations, these IPs could be potentially untrustworthy [2, 19, 65, 70, 71]. This

© Springer International Publishing AG, part of Springer Nature 2019

S. Ray et al., *Security Policy in System-on-Chip Designs*,

https://doi.org/10.1007/978-3-319-93464-8_4

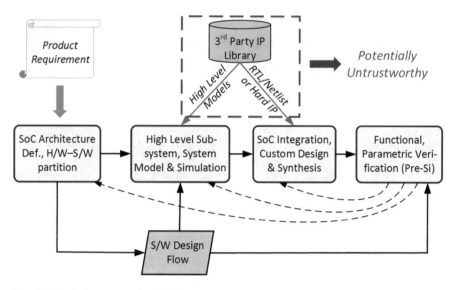

Fig. 4.1 Typical representative SoC front end (until fabrication ready) design flow

includes the possibility of a malicious vendor or even a rogue employee in one of
the vendor businesses inserting stealthy malicious logic or covert backdoor channels
into the design [54, 71, 76]. Commonly referred to as *hardware or firmware Trojan*,
these can make the system fail during critical system operation or leak sensitive
information from the system to an unauthorized agent. Like malicious insertions
possible in an IC at an untrustworthy foundry [11], these IP level Trojans could be
implemented so that they function or perform a malicious activity outside the design
specification boundary (additional to meeting functional, parametric specifications)
and/or utilizing rare system or internal IP event/s to trigger the rogue operation
within specifications (shown in Fig. 4.2), making them extremely difficult to detect
using generic functional and structural testing [16, 53]. Moreover, hardware Trojans
may also be activated by S/W-based triggers or their payload controlled by S/W,
which allows flexibility to an attacker to dynamically change untrustworthy Trojan
behavior [19, 38].

Some examples of representative hardware or even firmware (often supplied by
IP designer) level IP "Trojans" in this context in terms of their effect include a
processor core maliciously sending security critical register values to I/O memory
(or debug port), conditionally using unused opcodes or in addition to existing
instruction function, a memory controller generating an additional bank read/write
request for a load/store to a specific address range, or even a crypto-processor sending on-chip keys during encryption out to memory/system interface in concurrent
shadow mode. Similarly, as an example of such a threat in the communication
fabric, a router in a network-on-chip may conditionally (e.g., depending on source
and destination addresses and packet content) send a data packet to both the
destination address and a chosen address to create a path for potential leakage.

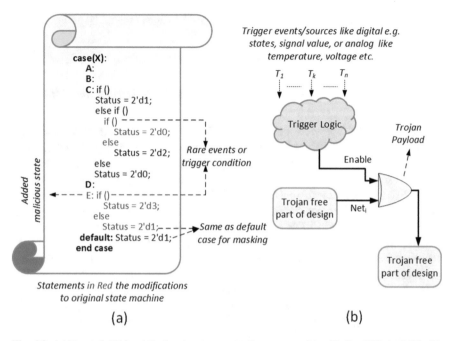

Fig. 4.2 (**a**) Example IP-level Trojans in a representative state machine Verilog RTL (soft IP); (**b**) logic-level representation of a sample Trojan Model

At the system level, a rogue action by an untrustworthy IP core may affect, influence, or trick other SoC components directly/indirectly to leak security critical information or function in a way so as to cause system malfunctions. Apart from an intent of malice, due to lack of strict design/verification rules, regulations, or cost limitations in certain licensing scenarios, unintentional vulnerabilities may also percolate in with the third-party IP block. The numerous instances of bugs found in a design on field, some of which could be security critical, attest to a chance of high probability of these inadvertent vulnerabilities [71] escaping test. Critical scenarios that have been backtracked to such loopholes include easy accessibility into secure grade FPGA design through the unprotected test port [66], intermediate values of round keys in encryption being reflected indirectly on system bus, etc. Overall, for a vulnerability, whether it be malicious or unintentional or whether the attack outcome depends on the particular business model, etc., it can potentially be exploited by adversaries of different capabilities to cause erroneous system functions or leak security critical data. This could be extremely critical in mission critical applications, e.g., defense, automotive as well for financial institutions which could be the target of the adversaries of the highest capability in terms of available resources, technical know-how, and financial backing. We note that as compared to the previous studies regarding analyzing Trojan effects in particular standalone ASICS, IP cores, processors, etc., and methods for verifying these designs for trustworthiness [10, 11], in this chapter, our focus is on analysis of system- or platform-level security vulnerabilities in the SoC designs created by integration of these untrusted IPs.

4.3 System-Level Security Issues Caused by Untrusted IPs

We start with an overview of system-level security issues caused by untrusted IPs to highlight the distinction between our work and related work on IP-level hardware Trojans. Table 4.1 lists the key distinctions between general trends in Trojan research and scope of this work. There have been previous studies on the feasibility and potency of malicious modifications in some ASIC designs and in some instances processor cores [11, 38, 63] in isolation. This includes inverting internal logic nodes via XOR gate and rarely activated combinational trigger logic as well as sequential Trojans or "Time Bombs" which can reset or disable the IP operation based on a counter reaching a chosen threshold (starting from reset) or after a particular sequence of input values. However, these have not involved analysis of untrustworthy IPs in the context of a system or platform such as an SoC where multiple hardware-, S/W-, or firmware-controlled IP modules perform dedicated roles and communicate with each other to perform system-level functions. For an SoC design, Trojans or malicious logic in an IP often visibly affect the overall system function rather than the IP core itself. In an SoC, with respect to a Trojan, we are concerned with these effects on overall system function rather than each IP core itself. For example, a malicious IP may send spurious communications to other IPs resulting in leakage of sensitive information, data corruption, or denial-of-service of the entire system. A critical problem with such system-level Trojans in an IP $\mathscr{A}$ is that their effect can only be observed in an overall system context—typically as a direct malicious effect from another IP $\mathscr{B}$—and may remain undiscovered in standalone functional or IP-trust validation of $\mathscr{A}$. On the other hand, due to scalability reasons, system-level validation (of both functionality and security assurance) with real use cases can be exercised only on executions using fabricated silicon [50, 75]. However, aggressive time-to-market requirements imply a limited window for post-silicon validation before the product goes on-field, resulting in potential escapes to shipped systems. Hence, potential system-level Trojan attacks can easily slip through even in the presence of efficient static IP-trust verification

Table 4.1 Current trends in Trojan research and scope of this work

Existing research on untrusted IP	Scope of the proposed work
Analyze feasibility and effect of Trojans in ASICs and processor cores	Analyze effect of IP-level Trojans at the SoC or system level
Explore static IP-trust verification techniques for Trojan detection	Run-time detection of potentially suspicious IP activity affecting system
Run-time methods that do not address error correction or recovery	IP-Trust aware security policies analyze and assert appropriate security controls

techniques. System- or platform-level attacks are being generally defined here as attacks which involve utilization of other IPs or system components by the rogue IP core to trigger and propagate the effect of the internal malicious activity at the system level and thereby compromise SoC security in some way. The objective of this work is to provide architectural support for on-field detection (and mitigation) of system-level Trojan threats.

System-Level Trojans vs. Trojans in a Malicious IP

While there is no standard, universally agreed taxonomy, most industrial SoC integration teams have developed a categorization based on the analysis of system-usage scenarios and protection requirements at different phases of system execution. In particular, an untrustworthy IP can typically affect system-level behavior through message communications, resource sharing, and control of operation and data flow. We can classify malicious behavior either in terms of the kind of threat introduced or in terms of the system-level impact of the Trojan. We can classify the rogue IP *threats* into four categories [51]: (1) *Interception*; (2) *Interruption*; (3) *Modification*; and (4) *Fabrication*. Correspondingly, in terms of behavior, malicious IPs can be characterized with the following taxonomy:

- *Passive Reader*: An IP that illegally reads/collects secret information meant for other IPs.
- *Modifier*: An IP that maliciously changes communication/message content between two IPs.
- *Diverter*: An IP that diverts a message/information between two IPs to a third IP.
- *Masquerader*: An IP that poses or disguises itself as some other component, in order to request service from or control the operation of other IP/s.

Unlike IP-level Trojans, the taxonomies above characterize system-level Trojans by *impact* rather than their design, implementation, or triggering characteristics, i.e., like the threat classification, the four categories above, all capture maliciousness as a "black box," i.e., they do not care exactly what kind of malicious logic/circuit has been inserted, but they simply look at the effect in the context of other IPs in the SoC. An upshot is that the taxonomy does not directly translate to a scheme of statically checking a design for system-level Trojans, and one must resort to validation of the run-time behavior of the system either through dynamic or formal analysis. Besides, the categories are not mutually exclusive, e.g., a passive reader may also act as a masquerader or a diverter may also read information before diversion and hence also a passive reader. Furthermore, as illustrated in Fig. 4.3, any adversary behavior can result in multiple threats, e.g., a masquerader can cause any of interception, interruption, or modification.

Below, we provide some examples of system-level Trojans arising from a sample of potentially malicious IPs. While the descriptions themselves are simplified for pedagogical reasons, they are inspired by realistic protection/mitigation strategies developed by security architects in typical industrial SoC design flows. We note however that they are only meant to be an illustrative sample; an exhaustive overview of the spectrum of vulnerabilities caused by untrusted IPs is beyond the scope of this work.

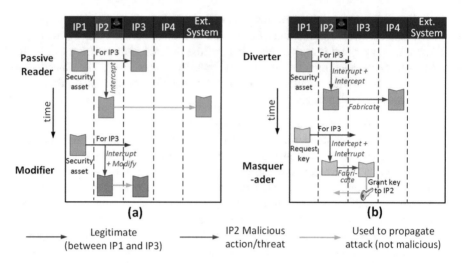

Fig. 4.3 Message diagram-level representation of untrustworthy IP being a (**a**) passive reader and modifier, (**b**) diverter and masquerader, along with the associated threats

1. *Untrusted Processor*: In a processor, the underlying control and/or data path may be slightly modified so that an existing instruction or opcode will perform something extra (potentially malicious), in addition to executing its normal functions as expected per the specifications. These changes may be in the hardware logic itself, at the microcode or firmware level or a combination of both. Along similar lines, the Trojan might be also in the form of a new instruction utilizing reserved opcodes. (These may also bypass detection provided that the typical security policy concerning the processor does not check for a golden set of instruction opcode fetches only.) To cause an intra-processor malicious activity to propagate outside and affect the system, an attacker may zone in on processor sites/events involved with system interaction. This would typically include the memory subsystem (including DMA based I/O devices) as well as test/debug interfaces. As examples of the additional functionality, a Trojan inside a processor might lead to a single memory access instruction resulting in two memory data access requests (apart from TLB) only when the address is within a desired range (which can be potentially configured from time to time to avoid detection). The additional request generated may be termed as the shadow memory operation [70]. In other words, for a simple RISC Load(LD)/Store(SW) architecture, a particular LD/SW instruction would lead to two LD/SW requests if the address is within a particular range. The underlying modification could be such that irrespective of the adversary software privilege level, the shadow mode memory access would always occur in the kernel mode (or the highest in system) with the system mode bit/s set in register (H/W), giving the shadow operation visibility to the complete system memory available to the processor (and potentially other IPs) as well as evade detection by generic privilege-based

access control security policies. This may be utilized to bring into the program window or tamper with security critical assets stored in the protected memory space. For the former scenario (load), the resultant secret may be written back to DMA memory space (and signal DMA controller for action) for the attacker chosen I/O device of choice and thus leaked. Thus, such an attack (leading finally to exposing a secret to an attacker) involves potentially multiple IP blocks (memory controller, DMA engines, device controller, etc.). With some knowledge of the final system address ranges in the SoC, the shadow load address arguments may be passed on accordingly (directly/indirectly) with the higher S/W stacks during execution. In terms of taxonomy, the untrusted processor core could be here classified as a passive reader, modifier (for a store), or even a masquerader.

Another example of this additional modification, as part of an existing or new instruction might be added malicious hardware logic to change the control flow of programs on demand, i.e., instead of using buffer overflow and other S/W attacks to change the control flow, one may think of this as hardware-based control flow attacks. The modification may be designed in a stealthy manner such that the instruction pointer (visible to programmer/debugger) might still be pointing towards legitimate instructions in sequence whereas in something analogous to a shadow mode, the underlying hardware is executing code from the malicious source of attacker's choice whose initial instruction could be a jump/branch (to avoid detectability of any kind). This added indirection would make it extremely difficult to detect this rogue action. A potential trigger for this attack may be a security critical program trying to access a secure key from on-chip secure flash that would be required for decryption/authentication and consequently control operations for a critical system. The adversarial malicious program may interrupt this flow thereby leading to denial-of-service (failure in mission critical system) or leak the secure key (stored in a processor register) to the attacker device of choice via DMA or test/debug port (attacker may have on-line access to debugger). With the key, an unauthorized agent can gain entry into the mission critical system and control/tamper it. Hence, different IP cores may be utilized by a small malicious modification in an untrustworthy processor to conduct a system-level attack. Here, the untrusted processor is a passive reader or modifier and/or even a diverter (potentially diverts keys from flash to I/O device). We note here that for the processor and the other remaining example untrusted IP types considered, we do not delve into possible implementation details of the particular backdoor or modification at the micro-architecture or logic level, but rather analyze it from a high-level, architecture-based behavioral abstraction.

2. *Untrusted Memory Controller*: Like a processor, the controller to the system memory, which typically comes in the form of a separate IP, may be untrust-worthy as well. It is a critical component as it governs the accesses to the memory, which is shared between different IPs and system components in an SoC. With memory-mapped input–output and DMA, the memory controller governs access to external devices as well. An example scenario may be a Trojan in the memory controller which tampers with the store values of only

a particular IP in certain trigger conditions. This specific IP could be, for example, a processor in a multiprocessor SoC, which controls the different sensor functionalities in a mobile system (like accelerometer, camera, microphone, temperature, keypad pressure sensors, health sensors, etc.). Tampering may be in the form of modification to data values in the request buffer before storage into the memory bank. This may cause malfunction in the control of these sensors and thereby hamper usability/availability or in extreme cases, such as temperature feedback control, can cause the entire system to overheat and fail. The tamper might also be in form of an extra shadow copy/write of an input key or signature (input via camera, microphone, or keypad) to attackers program space in the memory so that a program of the adversary's choice running in a different processor can later extract it. Hence, the memory controller could behave as a passive reader or modifier. Besides tamper of access requests, a hardware logic or often firmware modifications may be intentionally inserted to cause the memory access scheduling policy in the controller to ignore a particular IP's request on a certain trigger based on system context. As a result, the particular IP may be starved of resources to perform an operation, which can lead to a system failure in the form of erroneous outputs or denial-of-service. Here, the controller acts as a modifier, altering the scheduling policy.

3. *Untrusted Network-on-Chip*: Another shared system resource which is heavily used by the different IPs to communicate with each other to perform system-wide functions is the Network-on-Chip (NoC). NoCs are composed of multiple levels of routers or switches which direct the packets of information from the source to the intended destination. Typically, the full communication fabric or the NoC is designed as an infrastructure IP. We can imagine different scenarios of malicious modifications in the hardware logic and/or firmware of the routers that can cause widespread system-level effects. A particular case may be that for only preselected specific secure data packets or assets going from memory controller to a crypto-IP at system power up or down, a particular malicious router creates a copy and routes it as a packet to a device controller with external system access (along with the original intended destination, i.e., crypto-IP). This may expose the system in a certain security critical usage scenario to replay attacks, unauthorized access breaches, etc. Here, the router can be classified as a diverter and/or a passive reader. In a particular system context with specific real-time constraints, a Trojan in a router may be activated to increase the packet latency by altering the route schedule policy (e.g., from a shortest time to least link bandwidth consumption). This delay may result in a payload of a failed system request or function in the real-time system. The router acts as a modifier in such a scenario.

4. *Untrusted Device Controller*: Finally, along with the others, a controller to devices like USB, Bluetooth, Ethernet, modem, and other I/O components may be untrustworthy as well. As a simple malicious modification, a device controller (e.g., display controller), through insertions at the hardware logic or firmware, may modify bits of device data input under certain trigger events such as a particular authentication request (for log-in) in a program on the processor (would involve display controller operation for communicating request, thus

acting as trigger), that would require a user to input credentials. This intentional tamper would lead to denial-of-service attacks. Similarly, as device controllers are the link between the SoC and the external system, there could be many other possible attacks of denial-of-service types and/or fabricated device requests from the controller itself (not originating from any IP or program in processor) that may result in information leakage from other security critical components of the system (outside SoC, e.g., system flash/ROM). Hence, in the examples presented, a device controller can be classified as a modifier or passive reader and/or a masquerader.

4.4   SoC Security Architecture Resilient to Untrusted IP

In this work, we propose a security architecture that implements fine-grained, IP-Trust aware security policies at run time, which ensure reliable, secure operation of the SoC even in the presence of untrustworthy IP activity. The proposed framework detects potential undependable or suspicious behavior of untrustworthy IPs during execution and asserts appropriate security controls as defined by the policies. The architecture is an extension of the proposed security infrastructure "E-IIPS" [6], which presents a systematic, methodical approach towards implementation of diverse types of system-level security policies in an SoC, as described in Chap. 4. The generic security policies considered in [6] deals with the typical threats of attacks via the SoC to system interface and malicious S/W stacks executing on different IP cores. This work significantly enhances the E-IIPS architecture with necessary support to provide a disciplined, systematic, and scalable approach for addressing the threats of IP-level Trojans in SoCs, irrespective of the type of constituent IP/s.

4.4.1   Assumptions

As mentioned earlier, in the E-IIPS architecture, the malicious and/or unintentional hardware logic or firmware code may reside in the IP Core or the security wrappers, which are typically provided by IP providers, similar to P1500 test wrappers for boundary scan [17]. In the IP, we assume that apart from the standard, commonly used, and highly validated test and debug wrappers, a Trojan might be inserted anywhere within the internal control logic, data-path, and/or temporary storage locations. For the "smart" security wrappers capable of detecting different security critical IP events/data, apart from the frame generation logic and wrapper interfaces with SPC [6] and local DfD [7], which are relatively standard (in terms of high-level specification, behavior, and micro-architecture) across all IP types, the Trojan or unintentional vulnerability might be present undetected anywhere in the wrapper sub-components. As touched upon earlier, although two IPs or system components inside the SoC may be both untrustworthy, probabilistically, the event of them being malicious on the same system context is extremely low. Independence between different IP vendors and complexity of conducting coordinated attacks in the scenario of multiple IPs from the same vendor form the rationale behind this assumption. Hence, with respect to analyzing a potential defense against an untrustworthy IP (core + wrapper), the other interacting components in E-IIPS architecture like the DfD or debug trace macrocell, interacting IPs/system components, and communication fabric between the wrapper and the SPC are considered secure and trustworthy, i.e., the IP cannot collude with other components cohesively to conduct a potential system-level attack. The architecture of the debug framework is standardized for adoption in different SoC design scenarios. It is often an infrastructure IP belonging to the design house itself and hence trustworthy. These abovementioned assumptions are listed in Table 4.2. First, we present the solution for the untrustworthy security wrappers in isolation and consequently we deal with the problem of Trojans in the IP core. At the end, we integrate the two proposed approaches to enhance SoC robustness against the scenario of Trojans in any part of the IPs.

Table 4.2 Assumptions regarding trustworthiness of associated components in solution methodology with respect to an untrusted IP

Component in SoC	Trustworthiness assumption
IP core	Except standard test (P1500) and debug wrappers, all untrustworthy (incl. data path, control, and storage)
IP security wrapper	Except standard wrapper to SPC, local DfD interfaces and frame generation logic, all untrustworthy
Local Design-for-Debug (DfD)	Trustworthy
DfD to wrapper link	Trustworthy
SPC to DfD config. link	Trustworthy
Interacting IP wrapper, comm-unication link	Trustworthy
Wrapper to SPC communication link/fabric	Trustworthy
SPC	Trustworthy

4.4.2 Untrustworthy Security Wrappers

Among the typical wrapper sub-components, the event monitoring/extraction logic including the associated metadata storage may be highly susceptible to a malicious exploit. This is because they vary according to IP types and particular implementation. They also serve a critical role in the E-IIPS architecture in lending "intelligence" to the wrappers for detecting the necessary, critical IP events and can be directly used to launch an attack on the system. A malicious or unintentional vulnerability in the event detection hardware logic may cause the security wrapper to not report certain security critical events as and when they occur internal to the IP (False Negative), possible under some stealthy condition-based triggers. As a result, irrespective of the trustworthiness of the underlying hardware of the IP core, the SPC would be oblivious to any behavior/activity (rogue or trusted) originating inside the IP and potentially propagating to the system level. Hence, the system security state maintained by the SPC would not be updated and hence not reflective of the actual system context. This would result in incorrect security controls being applied, which can potentially lead to system-level confidentiality/integrity as well as availability-based attacks. Example scenarios might include:

1. A thread executing in a processor core requests a conditional jump to SoC-level system address (where security critical assets are typically stored during SoC execution) and the wrapper does not report it to SPC only when address is in a certain address range which could be something like the particular storage space of DRM keys or the process scheduling block in kernel memory.
2. The program on a processor requests memory controller for sensitive data to output to a particular device channel via a DMA engine on platform and the memory controller/DMA (depending on policy implementation) wrapper does not report it to SPC only in the scenario of a particular channel (specific device) of interest.

For the former, suppose if the memory controller security wrapper is not designed for verifying accesses via this processor (may be SoC designer implements the corresponding check only in processor according to specification), then the malicious wrapper can lead to confidentiality/integrity/availability attacks. Similarly, in the latter case, in a particular scenario of design of processor, DMA engine, and memory controller wrapper, such exploits may present significant opportunities for leakage or tamper of security critical assets. On the other hand, the reverse (False Positive) may also occur with respect to a malicious wrapper behavior, i.e., the wrapper fabricates and reports particular security critical events to SPC, possibly under some trigger conditions, when actually they do not occur at all in the IP operation/execution. Such a scenario could be equally stealthy in terms of potential effects at the system level. An example of this, which could potentially lead to availability attacks includes:

1. A bus controller or a router in a network-on-chip falsely reports to the SPC that IP "A" has frequently requested access to IP "B" configuration registers or its local memory data under a particular trigger condition (may be when IP "A" starts a security critical operation as detected by a particular communication request pattern of IP "A," which goes through this router). The SPC, according to implemented policy, disables all IP A communication requests for the current execution run leading to availability attacks and potentially system failure.

It is true that any malicious insertion or modification or for that matter a bug in an IP security wrapper would be comparatively easier to detect by validation, as compared to those in IP core designs, due to the relatively simpler design specification of the security wrapper (in contrast to main IP functionality). Besides, an adversary during IP design typically does not know the actual SoC-level security policies governing the IP. It may so happen that the inserted Trojan is inadvertently disabled during wrapper configuration by the SPC or the policy is just implemented in such a way that the event that the Trojan masquerades is not directly used to modify any security controls. In such scenarios, the Trojan trigger or payload may not be realized at all. But, at the same time, to an attacker, a Trojan in the security wrapper is easier to design/insert, compared to IP core-level Trojans as well as it offers the adversary a potentially easier and direct way to conduct a potential attack at the system level. An attacker's job to propagate the effect of the Trojan to the IP interfaces and consequently to the system is much simpler in this scenario. Hence, malicious logic in a wrapper is definitely a security vulnerability in the scenario of incomplete trust coverage via functional/structural tests. We propose a solution to detect the untrustworthy security wrapper action at run time and thereby protect the security and reliability of SoC operations.

4.4.2.1 Solution Methodology

The proposed solution is simple. It is based on detection of untrustworthy security wrapper action by comparing and verifying it with a trusted monitor at run time. The trusted monitor on die in this case is the DfD module or trace macrocell, local to the IP as illustrated in Fig. 4.4a. As required for observability and some controllability during post-Si validation, on-field tests as well as patch/upgrades, a typical macro-cell [22] for a processor (as an example) in an SoC implements standard instruction and data value/address/range comparators, condition code/status flags match check, performance counters, event sequencers, logical event combinations, etc. Along similar lines, for different IP types, the corresponding DfD modules incorporate adequate resources, which can be utilized to detect almost all of the security critical IP events required for policy implementation [7]. Moreover, these local trace macrocells would potentially already be utilized to verify the wrappers during post-Si validation. Utilizing this to repurpose debug instruments for security, the central engine, i.e., SPC would configure the DfD module, local to the untrustworthy IP of interest at power up to detect part or all of the security critical events at run time. The

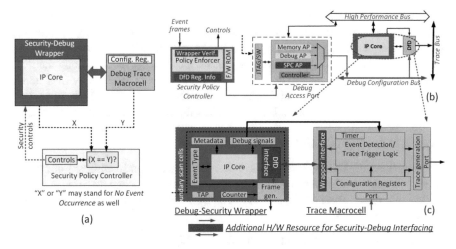

Fig. 4.4 (**a**) Cross-verification based proposed methodology to detect untrusted wrappers; (**b**) modifications required for repurposing DfD for security policies in SoC; and (**c**) zoomed view of the additions in IP security wrapper and corresponding DfD

number of wrapper events to be verified is chosen by the SoC designer depending on the degree of untrustworthiness of IP, system execution contexts, etc., and mapped to security policy functions in the SPC. For example, during security sensitive boot process, the DfD might be configured to detect all expected wrapper events, whereas during normal mode, a subset might be required to be monitored. For a security wrapper or trace macrocell communicated security critical event, the SPC compares the other's response (may be the same event, wrong event, or no event frame at all) and asserts appropriate security controls as determined by the policy. Here, an important point to note is that the previous chapter proposed the use of existing local DfDs (after configuration) to share the load of the wrapper by detecting some of the critical IP events at run time, thereby reducing the hardware resources of wrapper and making them lightweight. However, this requires custom design of the wrappers from the IP provider's part to adjust to different debug capabilities and design choices in a particular SoC implementation. This may lead to some increased design complexity as well as time-to-market. In our proposed method, we assume that the wrappers are not touched. It would constitute the corresponding debug interfaces [7] as specified by the SoC designer (for proposed verification), as shown in Fig. 4.4. During execution, due to power/energy constraints in particular use cases, the monitoring may be switched off as well.

The verification methodology is derived from N-version programming in software [11] that generates and executes multiple functionally equivalent versions of the same program to achieve high reliability in the presence of software faults. Here, only two independent sources are involved, one being trusted and the other untrustworthy. Potentially, the security wrapper of an interacting IP may be another independent source for event verification of untrusted wrapper of interest, as the same event context or sequence may be designed to be detected by both wrappers, depending on the particular implementation. For example, an IP's (e.g., processor) particular request to the memory may not be reported to SPC by its wrapper, but the memory controller's wrapper reports it when it receives the request along the flow of operation. Hence, a malicious IP wrapper action may be detected by SPC under scenarios when the actual event sequence extends beyond the IP and the interacting IP wrapper can detect it as well. Due to its rather limited scope, we do not consider it as part of the solution in this work. Note, however, that repurposing DfD for security requires addressing the following trade-offs.

Transparency to Debug Use Cases Post-silicon debug and validation are themselves critical activities performed under highly aggressive schedules. It is therefore critical that repurposing the DfD does not interfere or "compete" with debug usages of the same hardware.

Maintaining Power-Performance Profile On-chip instrumentation, and in particular the design of on-chip debug communication fabric, is optimized for energy and performance in usages related to debug. For example, since debug traffic is typically "bursty," it is possible to incur low penalty in power consumption even with a high-bandwidth fabric by power-gating the fabric components during normal execution; while repurposing the same infrastructure, one must ensure that the power/energy profile is not significantly disrupted by the new usages.

Acceptable Overhead for Interfacing Hardware The required hardware resources for interfacing with the DfD and configuring it for security needs should be minimal.

4.4.2.2   Implementation Details

For reusing DfD modules to detect security critical events for verification of the corresponding security wrapper, the SPC needs to have an interface with the debug infrastructure to configure the DfD module with corresponding values. As required in [7], the required modification in a typical debug access port [14], as illustrated in Fig. 4.4b, involves the addition of an SPC access port as well as possible modifications in the logic controlling accesses to the configuration bus. These typically constitute minimal hardware overhead [7]. With the DOT of the different IPs as well as the system execution context defining the wrapper events to be verified, the corresponding configuration register values of the associated trace macrocell are stored as policy arguments inside the SPC as seen in Fig. 4.4b. A hardware- or S/W-based event comparison is present as part of the policy enforcer. To avoid the IP provider in getting to know which specific security critical events would be monitored (and potentially insert Trojans accordingly) in the SoC and utilize it maliciously, the SoC designer (rather than the IP Provider) figures out the corresponding DfD configuration register values from the debug programming model [14]. This practice incurs some small additional design effort. Standard trace macrocells typically incorporate logic resources to detect all of the security critical IP events required for policy implementation. Hence, most of them can be configured at power up/boot time to be detected by the DfD block during execution. A few may be required to be programmed by the SPC only at run time, based on the policy requirements. Hence, apart from power up, the configuration bus is mostly OFF (power gated) and hence saves leakage power/energy. In certain rare scenarios, where there is less configuration register space for programming DfDs to detect all sorts of events simultaneously, a few new registers may be added to the debug configuration address space to trigger concurrent event monitoring.

Noninterference with debug usage requires transmission of security data from DfD to SPC via a separate port/link (instead of repurposing the debug trace port and the trace communication fabric, which might incur significant power/energy overhead at run time). This port/link would be triggered accordingly by the event detection logic in DfD. As in the usage scenario of [7], we interface the DfD with the corresponding IP security wrapper and send the DfD detected event/s through the existing active frame generation logic. Although the security wrapper is untrustworthy, as mentioned in Table 4.2, the DfD interface in wrapper and frame generation logic, which are mostly standard across wrappers of all IP types, are assumed trusted. With the wrapper to SPC communication link also considered trusted, the reuse of wrapper trusted resources to transport DfD detected event frames to the SPC would not give rise to any new security concerns and, at the same time, aid in meeting debug repurposing constraints. A close up illustration of a representative security wrapper, DfD module, and the additional resources required for their interfacing is provided in Fig. 4.4c. As a particular implementation, for the SPC to uniquely identify the DfD detected events from frames, the current value of the corresponding configuration register would be sent as part of the DfD frame. Besides, the implementation details govern whether wrapper frames and DfD frames are sent to the SPC simultaneously (as part of bigger frame) or consecutively one after the other. Irrespective of this, if a mismatch in the two values (including no event reporting from one) in the policy enforcer triggers the SPC to apply necessary security controls as governed by the policies. This may include cases such as SPC noting event at which mismatch happened and accepting only DfD events for that context in future, continue dual verification to monitor the degree of untrustworthiness of IP, and/or getting all consequent possible events from the DfD for high frequency of mismatches. These security controls can just configure the frame generation logic inside the wrapper to select the source/s of information. Comparatively, the incurred power/energy overhead of a few active trace macrocells, as required by our method, is minimal in contrast to a full active debug operation. During the debug phase, where the particular trace macrocell may be required for other monitoring/controlling purposes, proactive security policies (e.g., disable untrusted IP interaction with a particular IP until validation completion, etc.) would be implemented in the SPC to avoid requirement of most of the fine-grained IP security critical events in making a decision. This would increase the probability of malicious logic in wrappers not triggering and/or realizing their payloads.

4.4.3 Untrustworthy IP Cores

As seen in the previous works on static trust verification of designs, untrustworthy IP cores is an extremely challenging problem in terms of finding an adequate solution to provide high-security coverage. We note here that our proposed run-time SoC protection mechanism against potential system-level effects of untrustworthy IPs is complementary to these existing static IP-trust validation techniques, which attempt mostly to detect malicious modifications and/or covert backdoor channels in designs via targeted test vectors or formal analysis. In contrast, the aim of our dynamic protection is to monitor and detect system (SoC)-level direct/indirect effects of IP-level Trojans (and bugs) at run time and apply necessary security controls, according to requirements of corresponding fine-grained IP-trust aware security policies. Although we do not claim complete coverage against the gamut of all possible untrustworthy IP core scenarios, the intention is to show that just like the scenario of SoC with underlying trusted IP hardware, where security policies defend against threats, mainly originating from malicious S/W stacks and SoC to system interface, the SoC designer can also implement policies to detect untrusted, undependable IP actions arising from Trojans in the design, and prevent any system-level compromise. At the same time, one can do so in a systematic, methodical fashion with some enhancements to the E-IIPS architecture. As opposed to an exact set of rules and regulations, the solution provides guidelines to SoC designers/integrators on an efficient approach towards solving untrustworthy IPs in SoC issue. We note here once again that we have assumed that there is no malicious collusion between IP cores to execute system-level attacks, i.e., we treat each IP as independent entity from viewpoint of untrustworthiness.

As mentioned earlier, for such third-party IPs, there is no golden RTL implementation or associated models available as templates to an SoC designer, apart from the high-level IP functional/architecture specifications (trusted as SoC designer/architect would typically provide it) and the SoC architecture (signifying that an IP's interface with other SoC components, IPs is known). Even if architecture is not explicitly specified for the IP by the SoC designer, high-level features like the number of pipeline stages, their overall functions, number of cache levels, presence of virtual memory or not for processors, and similarly for other IPs are mostly available and easy to validate by the SoC design house. The key observation here is that one can utilize only these high-level specification, IP interface-level information along with generic architecture-level rationale to verify correlations between specific, abstracted out, temporal events across different micro-architecture-level sub-components of an IP to detect potentially untrustworthy behavior that might affect the SoC operations. Typically, in a design like a trusted IP core, a functionally relevant operation, meaningful and visible to SoC components external to the IP, incorporates specific correlated, internal (to IP) events occurring temporally across multiple spatial micro-architecture-level IP sub-units, i.e., these sub-units interact in a specific rational, meaningful manner with each other to perform an activity or operation [70], relevant at the SoC level. The corresponding events are referred to here as "Micro-architecturally Correlated Events" (MCE). Typically, IP-level Trojans disrupt/affect in some way the correlation between these spatiotemporal events, i.e., in the presence of an activated Trojan, internal events bearing little or no correlation with recent one/s may appear. This is explained with examples as follows.

Example 1 For a typical in-order processor core with five pipeline stages, a memory subsystem request (or simply an LD/SW in a RISC core) would involve a typical MCE sequence, i.e., the decode stage deciphering the instruction to be an LD/SW, the execution unit calculating the address, and consequently the memory access stage generating the appropriate memory subsystem request. An example processor-level Trojan is one inside the memory access logic to conditionally generate a shadow LD/SW in additional to the normal one. Here, the correlation disruption is in the form of a single active memory instruction (after decode in instruction window) generating two data memory requests, which does not satisfy common architecture rationale for a simple RISC processor.

Example 2 A hardware Trojan in the instruction fetch stage of a processor, triggering a branch or jump (to potentially a malicious source) without any previous active branch/jump instruction or corresponding activity in the program counter (PC) select logic (apart from $PC + 4/8$) is an example of not satisfying correlation.

Example 3 In a typical memory controller architecture, for an example Trojan inside the request buffer, the earliest active request not being served under a FIFO-based scheduling policy or a random change to row buffer-based policy for a Trojan in the arbiter/scheduler (for just a normal request; no condition flags, etc.) are scenarios of uncorrelated spatiotemporal events.

Example 4 A rogue router conditionally generating a different or additional destination address and sending the packet there instead of or additional to the one in its active buffer also does not satisfy the MCE typical for a router from the viewpoint of high-level specification and common architecture-level rationale.

The major components of the proposed run-time framework that are required to be added to the "E-IIPS" architecture of Chaps. 4 and 5, to provide the abovementioned SoC security against untrustworthy IP designs, are described below. Utilizing these, a sequence of typical operations (across time) involved in detection of an untrustworthy IP action, that attempts to propagate to the system level is illustrated in Fig. 4.5a.

4.4.3.1 IP-Trust Aware Security Monitors

For targeting confidentiality (C) and integrity (I) attacks at the system level, the payload of the Trojan in the IP would be designed by an adversary to propagate the malicious action out of the IP. This would invariably involve output transactions with interacting IPs or SoC components. Detecting availability attacks is more challenging and is dealt later. For C/I attacks through these untrusted

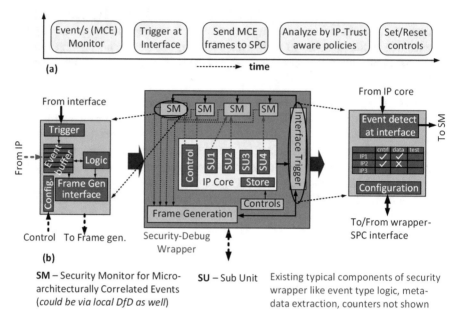

Fig. 4.5 (**a**) Typical sequence of events in detecting malicious action of IP in the proposed solution; (**b**) block diagram representation of architecture of enhanced IP-Trust aware security wrapper

IPs, we monitor high-level temporal events across the IP sub-units or MCE, that directly/indirectly affect or lead up to these output interactions. This is done by inserting "security monitors" inside the test-security wrapper to monitor and store a recent history of high-level events from strategic locations internal to the IP. An illustration of the architecture of a typical wrapper and associated security monitors is provided in Fig. 4.5b. From the high level, an IP core constitutes control logic to control the operations, data-path logic to perform the operations and local stores (e.g., register file, and reservation tables in processor) to temporarily store context. Typically, a Trojan would reside mostly in the control logic (including controlling the stores) and/or to some extent as modifications in the data-path hardware. Altering the control flow or adding new control states allows the attacker possibilities to conditionally trigger the designed Trojan and thereby ensure its stealthiness. As highlighted before, in terms of their expression/payload, one way to classify IP-level Trojans into two major categories is done as follows:

1. Performing covert functions or operation (usually conditionally triggered) in addition to what is specified as part of functional specification, operation manual, or normally expected from common architecture rationale, e.g., additional shadow loads/stores in processor, a router along with routing a packet from intended source to destination also changes its scheduling policy on conditional trigger from source and packet data (would affect later packets), memory

controller sending a particular request to two banks, etc. These are extremely difficult to detect as all intended functionalities are met.

2. Utilize rare event or event sequence conditions associated with IP to cause modifications to the existing or legitimate IP functions/state (not additional), e.g., in processor, a load/store to a different address than one intended in processor, an addition operation in the ALU instead of a shift/subtraction, etc., a packet in router misdirected to different destination, a memory controller following a row buffer-based policy instead of specified priority-based scheduling, in response to rare event triggers, etc.

The former might require some additional hardware/logic insertions as compared to the latter in some situations. But, it is more stealthy as both the factors of rareness of trigger conditions and operations in addition to intended specifications are involved. On the other hand, a miscalculation of profiling rare events at static/design time by attacker as compared to actual dynamic time traffic or events occurring in IP on interaction with environment may expose the latter to a higher chance of detectability. Typically, for most adversarial privilege levels, in case of the IP-level Trojan of the former type, the associated additional operation or function would have no correlation or be only minimally correlated with the IP-level context and recent IP operations/functions from a high-level abstraction. For example, the additional shadow load/store, originating from the decode stage or memory access sub-unit in a processor core is not preceded by a corresponding instruction fetch and/or decode and occurs randomly in the IP context (no correlation with recent past events). Similarly, for a particular memory controller context, a controller changing its scheduling policy in response to a set condition flag inside the scheduler hardware (for logging service efficiency, etc.) along with executing a simultaneous covert load/store without any active requests or active requests with that address/data pair is another example of uncorrelated events in the current context (schedule policy change) of the memory controller. For these examples, to potentially detect these malicious activities at run time, the proposed security monitors need to log events from the code fetch, decode, and memory access stages of processor as well as scheduler arbiter, request buffer, and controller–memory interface of the memory controller, respectively. For the latter scenario of IP-level Trojans, the above approach would work as well. Depending on the policy requirements, the security monitors may require to be at a finer information granularity and so do the policies checking the correlation. For example, where the malice expresses in the form of miscalculation of load address or addition in place of intended subtraction, instead of only checking whether a memory access or execution activity corresponds to recent decode action, the arguments also need to be taken into consideration, i.e., load address, type of ALU operation, etc.

Just like the scenario of categorization of IP types and standardization (to a large extent) of the security critical events as in Chap. 4, we broadly divide the IP types and illustrate a similar cluster of high-level IP-Trust aware security critical events or required "MCE" and associated policies for each type. These are enumerated in Table 4.3. Of course, depending on the degree of untrustworthiness of the IP

Table 4.3 Categorization of MCE and policies by IP types

IP	Ex. IPs	Ex. MCE list	Ex. associated policies
Proceessor core	GPP-, GPU-, μC-based device and system controllers (USB, video, power management, etc.)	Instr. fetch, decoded value, memory req. generation, cache access, interrupts/exceptions, prog. counter calculation logic	Decode preceded by corr. fetch; Fetch preceded by corr. prog. counter(PC) calc.; Mem. system access preceded by corr. decode; Main mem. access only on cache miss; PC jump only on corr. instr. or trap
Storage controller	Main memory controller cache controller flash, NVM control logic, DMA engines	New/next access in req. buffer, change in scheduler policy or priority of I/O channel (DMA), bank/col/row calc. by interface	Entry in req. buffer corr. to valid mem. req.; Req. scheduled corr. to prev. valid buffer entry; Req. scheduled acc. to policy by scheduler; DMA req. to mem. preceded by corres. μP grant
Comm. fabric unit	Router, switch, bus controller, bus bridge	New/next req. in buffer, next router along which packet sent, scheduling policy change, freq. scale in bus bridge	Entry in buffer corres. to valid comm. req.; Scheduled packet corres. to prev. valid buffer entry; Next destination chosen from current routing policy; Next bus transaction selected acc. to current priority

and the extent of IP and system-level validation performed on it, events may be added or removed from this MCE list or may be monitored selectively based on security critical contexts (like boot, debug, etc.) via configuration by the SPC, but this Table 4.3 just serves to provide a flavor and nature of these different events and policies to be analyzed according to the type of IP, to verify its trustworthiness at run time. For example, for models of hardware Trojan in the instruction fetch stage in a simple in-order pipelined processor, a typical high-level correlation check, that may be applicable under Jumps to system memory, is to verify the instruction/s fetched against the corresponding program counter calculated in last cycle or presence of recent asynchronous interrupts/exceptions according to status registers. For untrusted program counter logic, this may be subject to check once cycle back to correlation with the past instruction fetched/decoded. Our assumption is that an attacker would not insert these Trojans in multiple sub-units internal to the IP with a view of stealthiness or detectability. Hence, with most of the sub-components actually legitimate (not necessarily trustworthy to SoC designer), such correlation checks will not have to be subject to recursive checks to many cycles back and the malicious trigger can be detected at the bud or near its origin in space and time. For the example above, if the program counter logic is trusted from extensive functional/structural validation, the undependable activity may be caught in a single round of check. Similarly for a model of Trojan fabricating a memory request in controller to memory interface of an untrusted memory controller, under read/writes to shared protected memory range, a correlation check according to the IP-Trust policy "A particular memory access with address or address/data pairs must be preceded by its corresponding presence in the request buffer" would be able to detect any such trigger. Like before, such check may lead to recursive event verification in scenarios of untrusted request buffer control logic as well. In the business model, similar to wrapper specifications before, an SoC designer would additionally request the MCE requirements for the IP (to the IP provider) to be extracted out as hardware logic monitors as part of the wrappers. In many scenarios, the required events may be already detected by existing standard wrapper event type and metadata logic. These would reduce the incurred additional design modification and hardware overhead of the new monitors. In cases where SoC designer wishes to maintain greater control (and thereby trustworthiness) in trade-off with added design effort, the IP providers are requested to provide only the necessary signals to be brought out to IP interface and the SoC designer would insert these monitors during SoC integration. An illustration of potential sites, regions, or sub-units of representative IP cores of three different types, from where IP-Trust aware security critical events (intended MCE) may be extracted in a typical scenario, is provided in Fig. 4.6a–c.

With respect to IP providers providing the MCE monitors in wrapper, their trustworthiness is again of utmost importance in proceeding ahead with these run-time security checks. Similar to techniques mentioned in the untrusted security wrapper section, the integrity of these monitors can be easily verified at run time using appropriately configured local debug trace macrocells (which would detect these MCE for different cases as they are required for validation/debug as well). Based on certain outcomes, the DfD trace macrocell may be consequently utilized to serve as these MCE monitors. Alternatively, taking advantage of the high extent of standardization of DfD components (and hence less potential for untrustworthiness), these trace macrocells may be used by the SoC designer from the very beginning

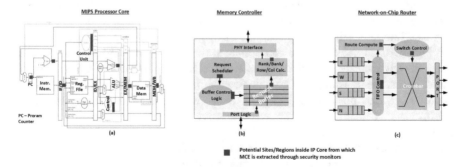

Fig. 4.6 Potential sites in the IP design for insertion of IP-Trust aware security monitors in (**a**) MIPS processor core, (**b**) representative memory controller, and (**c**) NoC router

to detect these critical IP-Trust aware events at run time. With appropriate interfaces present/inserted in the trace macrocells as mentioned in malicious wrapper solution, the SoC designer would have to decipher the corresponding trace cell configuration register values from the debug programming model. An illustration of the architecture of a typical wrapper and associated security monitors is provided in Fig. 4.5b in a typical case of the wrapper extracting the required MCE. The typical wrapper components are not shown here. As seen in Fig. 4.5b, the input to the security monitors (SM) is from control logic and/or data path sub-units of the IP core. The buffer inside SM stores the recent log of temporal events for the corresponding logic/stage. The SMs are triggered by the "Interface Trigger" logic to send the event logs to the wrapper frame generation (for communicating to SPC) on specific interface events of interest (described below). Based on validation/test coverage, degree of untrustworthiness of the IP as well as security requirements during different execution phases (coarse grained like boot, execution, power down, debug, etc.), the applicable security monitors are appropriately configured by the SPC at boot time.

4.4.3.2   IP-Trust Aware Interface Triggers

In the work, our goal is not to perform verification of all events of interest of the untrustworthy IP at all times according to intended high-level specification, i.e., IP functional validation for any potential faults or bugs is not the aim. Our proposed solution is based on the principle of isolating undependable or malicious activities, arising from Trojans or unintended vulnerabilities, to within the IP core itself. This leads to interacting SoC components not being affected directly or indirectly by the malicious action propagation and hence no compromise in system-level security, which is the high-level goal here. Hence, one would desire to perform this event correlation verification according to IP-Trust aware high-level policies, only when the corresponding IP attempts to communicate with the other SoC components through its output interface. As mentioned before, these attempts would be detected by the "Interface Trigger" logic, inserted as part of the security wrapper as shown in Fig. 4.5b. On detection, the appropriate security monitors are triggered to send their recent security critical event logs to the SPC for verification. However, in most cases, the constraints of communication bandwidth, SPC execution resource constraints, and power/energy profiles along with requirements of system-level performance demand for an optimized, selective scheme on which these events are subject to verification, i.e., trigger only on certain interface actions/events that may be configured inside the Interface Trigger logic. As mentioned in [70], the input interface of any general hardware module or IP core can be categorized into four types of signals:

1. *Control*—Controls the operation or function of the module, e.g., Valids, Status, Commands, etc.

2. *Data*—Provides the data or information on which the operations or functions are performed
3. *Test*—Constitutes signals required for putting the core in Test mode as well as for testing the module, e.g., scan chain inputs
4. *Global*—Constitutes signals like clock, reset, etc.

Similar to Trojan's effects inside control logic and data paths of untrustworthy IP cores, from the point of view of affecting the system-level security, the control and data inputs at the interfaces of the interacting IPs would be mainly influenced by potential rogue activity of the corresponding untrusted IP-level Trojan. Test and global signals are generally limited in number for a module and any influence on them due to Trojan has a much higher detection probability during validation. Hence, it is assumed that they are typically not utilized by attackers to propagate payload at the system level. For each IP or component interacting with the untrustworthy IP in the SoC, the SoC designer figures out the type of inter-IP communicating input signals (data, control, etc.) and analyzes the security criticality of the control and data inputs, if applicable, from system- and subsystem-level simulation. The corresponding interface trigger conditions or events such as "all activity for these specific control and data signal inputs at boot" or "specific events at these particular control signals at execution time," etc., are formulated and configured by the SPC in the interface trigger logic at power up. For example, in a particular usage scenario for an untrustworthy processor communicating with the memory system, all control (read/write) and data (address, data) input may be critical at boot, whereas during normal execution, along with control, only address in secure protected range is critical with respect to potential system-level security compromise. Similarly, for the same processor controlling a crypto-IP, both control (mode, enable, round status, etc.) and data (e.g., keys) are typically critical at boot (as keys are transferred to crypto-IP secure storage at that time) and only control inputs are security relevant during normal execution. A typical interface trigger logic has outputs to all the security monitors as shown in Fig. 4.5b. On detection of the corresponding interface-level event, the monitors are all triggered to communicate their recent event logs to the frame generation logic. These will consequently be sent to SPC for proper correlation validation. With regard to resource constraints and performance requirements, the frame-based interface may be made wider as compared to before, to allow for higher throughput of these events. This depends on the maximum number of events to be sent, any real-time requirements, etc., and hence is implementation dependent. Depending on whether any undependable or rogue activity has been detected or not, the interface triggers may also be configured by the SPC at run time to tighten or relax conditions. Along with trigger event selection, the corresponding interface logic also constitutes proactive controls that are programmed/controlled by the SPC during boot/run time according to requirements of IP-Trust aware policies (see below). For example, for an untrusted IP1 interacting with IP2, the corresponding security policies may require that the communication is proactively disabled until the critical events are verified to satisfy correlation. In another case, the policies may dictate the inter-IP communication and the verification of events by SPC to occur in parallel and if correlation checks fail, then roll back any state changes if applicable. Selection of IP-Trust aware security policies are described below:

With respect to the untrustworthy IP of interest, apart from these output interface activity acting as triggers for verification, the input control/data stream originating from the interacting IPs also may play a significant role in determining the trigger conditions and/or which events to monitor for validation in this scenario. In other words, for a security critical IP, whose control and/or data signals feed into the untrustworthy IP for some operations/computations, it is typically important to take into account the source of these control/data inputs in determining the granularity of event monitoring as well as interface triggers. For example, in a particular system context, when suppose a crypto-core (separate IP which is typically security critical as it may store security assets) communicates data (e.g., keys) to the untrusted processor for computation, any kind of output control/data activity directly/indirectly related to the crypto-data should serve as trigger conditions for validation, before it exits the untrusted IP. So, a tag, which could be as simple as a bit determining security critical or not, should be associated with the inputs from communicating IPs at the interface as shown schematically in Fig. 4.7. This tag bit/s propagate through the IP to security monitors and interface trigger, for all interactions inside the IP directly/indirectly associated with the associated critical data/control. Hence, the tag can be used for selecting which events to trigger on, which events to validate as well as what particular security controls should be applied at the interface trigger logic by the SPC during verification. A typical list of output interface trigger conditions (to verify IP-Trust aware events) in an untrustworthy processor for a particular scenario, with interface with main memory controller, DMA engines, flash controller as well as crypto-IP and DSP accelerators, is provided in Table 4.4. Here, along with event configuration in the "Interface Triggers," we assume that some of the system-level parameters like secure range of memory as well as specific security critical devices/channels are also programmed in there by SPC at power up (stored in SPC by SoC designer). From this representative scenario, we observe that the particular interface trigger events typically depend on various factors such as security tag of the

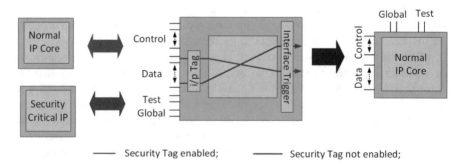

Fig. 4.7 Input tags associated with input data/control streams to untrustworthy IP, depending on the security criticality of the interacting IP

Table 4.4 Representative interface triggers for an untrustworthy processor

Interacting IP	Security critical? (tag)	Example o/p interface trigger
Memory controller (I/O)	Partially (RD/WR from/to secure memory)	If secure tag or boot phase, all control/data, else only RD/WR request to secure memory
DMA engine (I/O)	Partially (only I/O from/to specific devices)	If secure tag or boot phase, all control/data, else only I/O requests to specific devices
Flash controller (I)	Fully (flash contains bootcode)	NA (authenticated upgrade (o/p) only in special debug mode)
Crypto-IP (I/O)	Fully	All control and data at all times
DSP Accel. (I/O)	No	Control/data only at boot

outgoing transaction, boot or normal mode/phase of the system as well as security criticality of interacting IP.

Availability attacks via denial-of-service do not always require the propagation at the system level and thereby interaction with other SoC components, e.g., an IP1-like memory controller not responding to a request from IP2 under certain conditional triggers, may lead to system failure or compromise depending on the effects of the action on IP2. Hence, these are more difficult to handle as compared to the scenario of system-level confidentiality/integrity attacks that have been the major focus in this work. We present some directions or ideas along which an SoC designer might focus on to provide security against such denial-of-service threats. Similar to tags used to notify/identify security critical IP inputs and any associated direct/indirect operations on them, similar metadata scheme may also be added at the input interface to denote, for example, "Action Required" or "Some Response Required" for specific IP requests according to requirements of corresponding liveness policies programmed in the SPC. If the response (could be request grant or deny) is not received in the o/p interface within an estimated time or system context, the security monitors are triggered to send the event logs to the SPC via the frame generation logic for appropriate verification and action.

4.4.3.3 IP-Trust Aware Security Policies

IP-Trust aware security policies in the SPC dictate what and between which micro-architectural events the correlation checks should be performed and the applicable security controls at the interfaces of the untrustworthy IP/s before and after the verification of the IP behavior. Similar to the case of generic policies, they are programmed by the SoC designer as firmware in the instruction memory (could be flash, ROM, etc.) of the policy enforcer inside the SPC. Examples of typical correlation checks that capture the high-level (abstracted out) intended/golden behavior or operation of untrustworthy IP cores of different types are enumerated in Table 4.3. These intended high-level behaviors, involving interaction between different IP sub-units, are stored in the SPC. As mentioned above, apart from the verification of the event correlation, these policies also govern the security controls in the interface trigger logic of the corresponding untrustworthy IP wrappers. What values to apply to these controls at run time before any verification in the current context are influenced by a host of factors including degree of untrustworthiness (DOT) of the IP core, security criticality of the interacting IP in the current system, any real-time availability or performance requirements as well as past history of events from that IP. Depending on security criticality and or resource availability/performance constraints, two scenarios may typically arise for the untrustworthy IP:

- Disable all interface actions/activities until verification of the recent micro-architecture-level events for intended correlation by the SPC. This is a safety first approach which could be applicable for highly security critical interacting IPs, during the system boot phase where many security assets are shared between IPs and/or where the untrustworthy third-party IP has a high DOT.
- Allow the interface activities to take place, thereby allowing the IP action to propagate to other SoC components. At the same time, events sent from security monitors are verified for correlation. Apart from the scenarios mentioned in former case, these would be the method typically followed during normal execution. In case of undependability detected, the system may be rolled back by the SPC if applicable and possible. This might require maintaining of some check points by the SPC.

Moreover, there could be potential choices as to when the SPC would analyze the sent events. These could be critical for performance reasons in scenario 1 (described above) or preventing any security compromises in system in scenario 2. The straight cut solution would be for the SPC doing the correlation checks at the instant according to the programmed policies. Depending on the estimation of analysis or execution requirements at design time, a separate analysis engine could be added to the generic policy enforcer (which analyzes normal security threats from S/W stacks, system interface, etc.) to perform the IP-Trust aware verification separately. This could be just the addition of another micro-controller core inside the SPC. This is highly implementation dependent. The second solution regarding when the SPC should analyze events is adding support in the SPC to track untrustworthy IP action effects as it propagates through the system (using hardware-based tag bits) [52] and analyzing the events only when the effects are found to influence statically defined security critical assets/signals in other IP or SoC components (like configuration registers, stored keys, specific control signals, etc.). On one hand, this may come in handy in scenarios where the potentially untrusted IP actions are handled/controlled by the existing generic security policies in other SoC components/IPs and/or leads to no hampering of function, e.g., dropped packet in NoC or only latency increase, etc. A scenario of the former may be existing security policies leading to a memory controller rejecting an additional shadow load to secure memory generated in malicious processor. This saves execution resources of the SPC policy enforcer in terms of analysis of the events. But, on the other hand, similar to hardware-based tag support for S/W control flow integrity and information flow checks [18], hardware support is required in the SPC for tracking the potential IP action along the system. Resources are expended in determining the flow of action effects, etc. Hence, due to the two opposing constraints, the appropriate method to follow is dependent on implementation choices of the SoC designer. As mentioned before, the goal of the work is to provide general guidelines to the security community on possible methods or approaches to follow in order to tackle the untrustworthy IP problem in SoC. Finally, as part of the policies are also stored the conditions or events that need to be configured in the interface trigger logic of the untrustworthy IP wrapper. These detect the events related to the IP attempts to communicate with different interacting IP/s, on which to trigger the security monitors to send event logs for check. The SPC mainly configures these in the untrustworthy IP wrapper during boot. Sometimes, according to policy requirements for scenarios such as rogue, vulnerable activity detected, etc., these triggers may be required to be dynamically configured by the SPC to add or remove events.

Both solutions of untrusted wrapper and untrusted IP core provided above might be combined to provide integrated security in scenario of Trojans or vulnerabilities possible in any IP sub-unit or module (described in use case example below).

4.5 Use Case Analysis

In this section, in a particular usage scenario, we elaborate on how our proposed run-time solution can detect malicious IP activity originating from both security wrapper and IP core-level Trojans and prevent any compromise of the system security. Here, in this use case, we assume for illustrative purposes that the adversary inserts Trojans in both the wrapper and the core of the main memory controller and designs the same rare event-based trigger logic for both. The hypothetical attack model is as follows:

Information Leakage via Malicious Shadow Store to Memory of Specific Device For particular load/store request to protected region of memory during specific secure critical program execution (trigger), in the scenario of cache miss and associated eviction of dirty cache block to memory, the Trojan in controller–memory interface of the memory controller (MC) is triggered to cause a payload of an extra shadow store (along with store of evicted block) operation to the memory-mapped region of device of interest of the attacker. The attacker, through a controlled program on processor, can request an output to the corresponding device, which is performed through DMA. Hence, potential critical assets (e.g., keys, configuration register values, etc.) used by the original secure program/service may be potentially extracted. The trigger also leads to an activated malicious logic in the MC security wrapper to not communicate event logs to SPC (for correlation analysis) on trigger from interface.

Assumptions In this use case, we assume that the attacker (possible memory controller IP provider) can utilize the SoC designer configured secure memory range

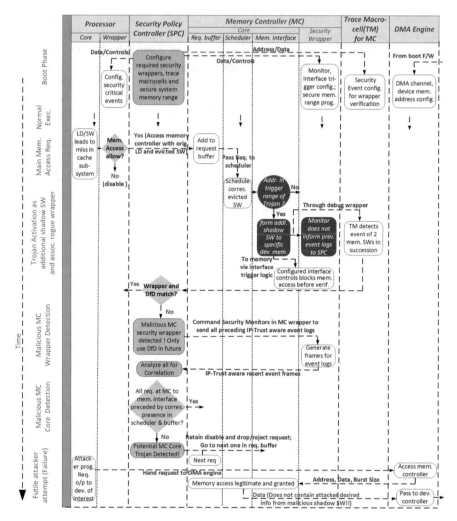

Fig. 4.8 Operation flow of the proposed solution for providing system-level protection in use case scenario of Trojan in security wrapper and core of main memory controller

(boot) in the memory controller (MC) security wrapper to act as trigger for the inserted Trojan in both the MC and wrapper. If not known at IP design details, the adversary can potentially utilize firmware (firmware) updates in MC scheduler to program the triggers (functional model similar to S/W-based trigger), assuming that firmware is involved in a typical MC design. Regarding the solution, as seen in the flow diagram of associated operations in Fig. 4.8, the processor core, memory controller IP, the trace-macrocell of the MC, DMA engine on platform along with the centralized SPC are involved in the system-level usage scenario. Although other IP cores could be independently malicious/vulnerable, in the context of the attack model, the memory controller is the only IP considered significantly untrustworthy here. Due to high DOT of the MC as well as security criticality of the shared memory, all accesses associated with (including evictions) secure memory region are disabled by the SPC according to security policies until the micro-architecture events are verified for necessary correlations. The MC DfD trace macrocell is utilized for corresponding untrusted security wrapper validation by the SPC. The memory controller is assumed to be constituting three major sub-components, i.e., request buffer, access scheduler, and controller–memory interface logic (which also computes appropriate bank/rank/row/column). Below is the sequence of steps that are performed along the system operation (as shown in Fig. 4.8) in detecting the Trojan/s and ensuring system security.

Flow of Operation

1. At the boot phase, the SPC configures the security monitors (if necessary) and the interface trigger logic of the MC security wrapper to program, for example, "all accesses associated with secure portion of the memory should be disabled until the preceding MC actions are verified for trustworthiness." The DfD trace macrocell associated with the MC is configured for part/all of the corresponding security wrapper events for verification. Besides the processor core, wrapper is configured with generic security critical events to monitor by SPC and the DMA engine by appropriate boot firmware.

2. During normal system-level execution, for a security critical program executing on the processor core, a cache miss occurs for an LD/SW to secure memory region. Associated with it is also a cache block eviction. The processor wrapper checks whether the memory access is even allowed according to privilege level of program by the necessary security policies. It passes the check and the accesses (along with evicted store) propagate to the memory controller where it is added to the request buffer.

3. Consequently, along with cache missed LD/SW, the evicted SW is scheduled by the arbiter (inside scheduler). This triggers the Trojan in the controller to memory interface to generate an additional shadow SW to device memory of attacker's interest (payload). Hence, both these requests would access the physical memory next. However, the configured interface trigger controls of the MC wrapper (by SPC policy) have disabled the controller interface until the verification of past MC events for trustworthiness. On this memory access attempt, the interface logic triggers the security wrapper to send the event logs for check to SPC. But, a triggered Trojan (same trigger delayed) in the security wrapper monitor prevents the SPC to be notified of these events. At the same time, the DfD trace macrocell detects the corresponding events of two pending stores to physical memory and informs the SPC through the debug–wrapper interface and frame generation logic. Receiving nothing from the security monitors, the SPC detects malicious or undependable MC wrapper. The SPC through wrapper–SPC interface commands the MC security monitors to send the event logs for verification.

4. Assuming the event logs to be legitimate (can be again verified by DfD, but not shown here), the SPC analyzes them for satisfying correlation according to MCE

of the memory controller IP core. Consequently, according to rule "All RD/WR accesses at controller–memory interface must be preceded by their presence in the scheduler sub-unit as well as the request buffer," the correlation disruption is caught and hence the Trojan action detected. According to policy requirements, the disable at the MC to memory interface is upheld and the corresponding accesses rejected/dropped. The next access from request buffer is served by the MC.

5. In due time (shortly afterwards as he/she does not want the corresponding device memory modified), the attacker, through running a program remotely/physically on the processor core, requests for data output from corresponding memory (as dictated by shadow SW address/es) to device of interest (may be network adapter, Bluetooth, Wi-Fi, etc.). However, with the attack thwarted by the proposed solution, the adversary fails in the attempt to extract potentially security critical information.

4.6 Overhead Analysis

In this section, we consider the hardware overheads incurred by the IP-Trust aware security monitors, inserted in the wrappers of different IP types, for varying Trojan coverage scenarios. We extend our developed SoC framework (used in Chaps. 4 and 5) in Verilog RTL, towards inserting these security monitors. The base SoC model schematic, showing the major IP components, is illustrated in Fig. 4.9a.

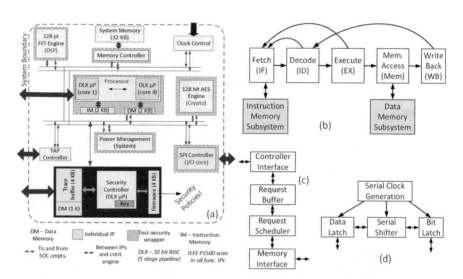

Fig. 4.9 (a) Block diagram schematic of the SoC framework; the internal sub-units of the (b) DLX processor, (c) representative memory controller and (d) SPI controller

4.6.1 Security Monitor Implementations

For incorporating the proposed security monitors, we considered three IP blocks of different types, e.g., the DLX processor core, the memory controller and the SPI controller. Their major sub-units are shown schematically in Fig. 4.9b–d. According to the methodology described in the architecture section for untrusted IP cores, security monitors are inserted in their wrappers at varying spatiotemporal granularities, for verifying different Trojan coverage scenarios. Table 4.5 lists the IP-level Trojans (in terms of their effect/payload), against which protection is ensured by these monitors and necessary IP-Trust aware security policies in the SPC. Interface trigger logic was incorporated within the wrappers too. For event communication to SPC, a separate, additional frame-based interface has been incorporated in the IP wrappers to avoid contention between the two traffic types (generic events and proposed MCE logs). We note that it is infeasible for the SPC to perform complete emulation of the high-level, intended IP operation. Hence, monitors must be placed at intermediate strategic locations in the IP as discussed in architecture section, to enhance efficiency of Trojan coverage. In scenario I of DLX Trojan coverage, monitors are placed at input of fetch, output of decode, and output of memory access stage, whereas in scenario II, monitors are additionally inserted at ALU outputs and write-back outputs. A malicious logic triggering a branch/jump for a particular unsatisfied condition (which implies that there should

Table 4.5 Different scenarios of Trojan (represented by payload) coverage by insertion of security monitors in three IP cores of our framework

Sec. monitor coverage scenarios	Ex. processor Trojan (payload) coverage	Ex. mem. controller Trojan (payload) coverage	Ex. SPI controller Trojan (payload) coverage
Scenario I	Addl. shadow (or hide) LD/SW by mem/ALU stage, wrong PC select logic, erroneous decode, wrong WB op. calc./propagate	Drop/Fabricate mem. req. by scheduler, wrong next req. select from policy (e.g., FIFO)	Error in serial clk gen. logic, wrong counter op. for external bit Tx/Rx
Scenario II	*All Prev.* + PC jump/branch calc. w/o corres. inst., mem./ WB stage perturbs ALU res./WB val., ALU perform wrong calc.	*All Prev.* + drop/fabricate req. by buffer, error in bank/row/col calc. at interface	*All Prev.* + data tamper at Tx/Rx shift, data latch at incorrect times
Scenario III	*All Prev.* altering mem. req. data/addr. wrong prog. branch T/NT + cache subsystem	NA	NA

be no branch/jump under normal operation) is detected using the ALU monitor in scenario II. Similarly, for the memory controller (MC), monitors at output of request buffer as well as scheduler can detect any malicious activity in scheduler hardware or firmware in scenario I. Additionally, storing the memory requests arriving at the input interface of MC allows one to detect Trojans in the buffer as well in scenario II. The framework has been functionally validated using ModelSim for typical use cases. All analysis for area/power have been performed using 32-nm technology library.

4.6.2   Results

The estimated area and power overheads for the inserted security monitors and interface triggers in the three scenarios of Trojan coverage for the processor core are shown in Table 4.6. We note that the overhead is calculated here with respect to the base design of DLX core with standard security wrapper. The overhead varies between a relatively small 6–11% across the coverage scenarios. Besides, increasing the additional frame interface width to 256 bits from 32 bits to transfer simultaneously the 8 temporal event logs stored in each monitor (8 chosen according to design details) incurs minimal additional overhead. In addition, a point to also highlight is that for significantly increased Trojan coverage from scenario II to III, the corresponding increase in hardware overhead is minimal. This signifies that one could potentially gain high run-time security against an untrusted IP at minimal hardware cost.

Similarly, the incurred hardware overheads for the memory controller and the SPI controller are listed in Table 4.7. For the memory controller, a 4-KB register-

Table 4.6 Area and power overhead of security monitors in processor IP (orig. area and power with 1 KB inst., data memory at 32 nm–352,405 μm^2, 12.56 mW)

Different security monitor scenarios	Die area overhead (%)	Power (active + leakage) overhead (%)
Case I (32 b o/p)	6.68	6.92
Case I (256 b o/p)	7.17	7.32
Case II	10.44	10.82
Case III	11.68	11.62

Table 4.7 Area and power overhead of security monitors in memory controller (MC) IP, and SPI controller IP (orig. area and power of MC and SPI with wrappers at 32 nm–629,433 μm^2, 13.81 mW; 5456 μm^2, 0.298 mW)

Different monitor scenarios	Memory controller		SPI controller	
	Area ovrhd. (%)	Active + leak pwr ovhd. (%)	Area ovrhd. (%)	Active + leak pwr ovhd. (%)
Case I	10.77	14.04	29.08	19.12
Case II	11.16	18.53	101.88	66.77

Table 4.8 Die area overhead (OVH) of security monitors (SMs) with maximum Trojan coverage wrt. to our SoC framework (area—13.1 × 10⁶), Apple A5 APL2498 (area—69.6 × 10⁶), and Intel Atom Z2520 (area—40.2 × 10⁶), all at 32-nm process technology

IP core	OVH (%) in model	OVH (%) in A5	OVH (%) in atom
Processor	0.31	0.059	0.1
Mem. control	0.543	0.103	0.175
SPI controller	0.043	0.008	0.014

based functional memory is also added to the base memory controller to calculate the security monitor overheads. Although for small IP designs, the area/power overhead could be significant with respect to the base (e.g., in scenario II of SPI controller), the overhead with respect to the full SoC die is insignificant for all IPs, as shown in Table 4.8. Along with our representative toy SoC, two commercial SoCs manufactured at 32 nm are also taken into consideration to calculate these approximate overheads. Note that the increase in the SPC overhead with respect to its base value, due to incorporation of additional IP interfaces for event logs, control signals, and all required IP-trust aware policies as firmware in instruction memory, has not been taken into account in this work. However, from the sample values in Table 4.8, we believe that even after incorporation of SPC overheads, the hardware overhead of this proposed architecture would be minimal with respect to the full SoC die. Perhaps more generally, our experiments show the design parameters and trade-offs that a security architect must analyze to deploy the framework in an industrial SoC design environment.

4.7 Conclusion

In this chapter, we have presented, for the first time to our knowledge, comprehensive analysis of trust issues at SoC level caused by untrusted IP blocks. We have also presented a novel architecture-level solution to achieve trusted SoC operation with untrusted IPs. With growing reliance on third-party IP blocks during SoC design process, untrusted IPs are rapidly becoming major security concerns for SoC manufacturers. Design-time IP-trust verification approaches proposed in the literature to date fail to provide high confidence in identifying Trojans of all forms and types as well as possible exploits of apparently benign design artifacts, such as the design-for-test infrastructure. The proposed architecture provides a relatively low-cost and robust defense against untrusted IPs. The architecture employs fine-grained IP-Trust aware security policies to detect and prevent malicious operation of an untrusted IP at the system level. It builds system trust by considering trustworthiness of minimal set of standard components (e.g., design-for-debug structure and a security policy checker), which are suitable for comprehensive trust verification. The proposed architecture is evaluated for diverse use cases, which proves its effectiveness for representative SoC designs. It is applicable, in general, to various types of SoC designs and is scalable to large complex SoCs with arbitrary number of IPs.

4.8   Bibliographic Notes

Malicious logic in IP design is difficult to identify by standard functional validation [16, 53]. In particular, Trojans are typically designed to be exercised by rare events under very specific execution corner cases that are difficult to excite in a functional validation environment. We illustrate examples of Trojans inside a simple, representative finite state machine (FSM) in Verilog RTL in Fig. 4.2a. State machine-based controllers are usually components of all IP cores such as processor, memory controller, bus control module, etc. Here, addition of a new, potentially stealthy state and modifications to an existing state of the variable "X," taking advantage of some example rare conditions, are highlighted. For a processor, this might represent the usage of reserved instruction set opcodes for rogue functions or perhaps additional operations for an existing instruction that can be used to indirectly leak data. For a memory controller, such a Trojan might lead to modification in memory request scheduling policies and thus potentially starvation of resources for particular IP/s. Figure 4.2b is a lower-level (logic/circuit level) representation of a typical Trojan Model in a design highlighting the notion of "triggers" and "payloads." Researchers have looked at developing testing methods aimed at static trust verification of IPs [5, 54, 72, 76]. Detecting Trojans in these third-party IP blocks is extremely challenging as there is typically no golden (Trojan-free) RTL for the IP, rather only the functional specifications. As a result, the number of possible ways to express a Trojan in the circuit is often unbounded and grows exponentially with design size. Although techniques like that in [5, 31, 72], based on probabilistically suspicious nets, nodes, region, or unused circuit detection and generation of optimized targeted test sets, are shown to be effective in particular instances, they suffer from significant false negatives/positives based on chosen test set, threshold parameters, design type, etc., and hence do not provide complete IP-trust assurance. Even with assumed combinational and sequential Trojan models, the Trojan coverage for relatively small (compared to today's designs) ISCAS benchmarks is on an average ~80–85% with suitable chosen parameters [13]. The efficiency of these techniques with respect to test time as well as Trojan coverage reduce with increasing sizes of modern designs. Along similar lines, although formal verification of designs for direct/indirect effect of potentially untrustworthy sources on security assets is effective for small designs [45, 54], they are just not scalable to most industry grade IP sizes used today. In modern day SoCs composed of hundreds of IPs of different sizes and types, the net untrustworthiness after static trust verification may involve a cumulative effect of the uncovered (by tests) trust issues in the IPs themselves. Hence, along with these static methods, execution time dynamic checks are required, that would serve as the last line of defense, i.e., operations of IPs with varying degrees of trust and their communications with other SoC components should be closely monitored at run time for potential malicious or undependable behavior, to ensure secure SoC execution. Here, security also includes reliability of operation of the SoC as Trojans may affect inter-IP control signals (which may not be considered formally as security assets), which in turn can cause system failures.

A few run-time monitoring techniques have also been proposed for detecting hardware Trojans, but most have been studied and analyzed in the context of processor (μP) cores [19, 70]. However, they primarily focus on standalone IP validation and are difficult to adapt or scale to Trojans affecting system-level behavior, influencing other IPs in the SoC design. Furthermore, these approaches do not address online mitigation of detected threats in a platform. We feel in order to reduce the significant incurred resource overhead due to the continuous monitoring support suggested in some solutions [19], an untrustworthy IP action should only be verified if it propagates to and attempts to modify system state or affect other system components and potentially security critical assets along the flow of operation. This is in accordance with the well-accepted security principle of isolating a malicious behavior locally within a rogue IP core or module, similar to the concept of enclaves, containers in S/W security [32]. Furthermore scrambling of inputs to untrusted units according to coverage test set may require interaction with random number generators, encryption engines at input and output interfaces of the IP [71]. Apart from H/W overhead, this might cause significant performance degradations of the system as well. Hence, the existing dynamic techniques are limited in their applicability at the SoC level.

A version of this research appeared in the IEEE Transactions on Information Forensics and Security [8].

Chapter 5
A Move Towards a Hardware Patch

5.1 Introduction

In Chap. 3, we briefly touched upon the possibility of hardware patching. In this chapter, we explore that idea more fully and examine how we can use the architectural framework developed so far as the basis of hardware patching.

First, why do we care about patching? One key reason is the rapid growth of Internet of Things (IoT)[23], which have caused the edge devices to give access to a tremendous amount of security assets every day. These assets must be protected from unauthorized or malicious access. On the other hand, these devices can stay long in the field, e.g., a car or a smart meter can have a field life of a decade or more. This is a long time! Security needs can change in the course of this period, in response to previously unanticipated use cases, vulnerabilities discovered while the device is in field, and even advances in device technology (e.g., if quantum computers become a reality then many of the cryptographic protections in current devices would be compromised).

Traditionally, vulnerabilities discovered in-field have been addressed through software and firmware patches. However, such patches are "cludgy," awkward, and in many case difficult to apply. Furthermore, the ability to reconfigure systems in-field through software and firmware requires that bulk of the critical functionality of the system be implemented in software/firmware, not hardware. This has significant implications on energy and performance of the system. All other things remaining equal, software implementation is typically more energy-hungry and lower in performance than the hardware implementation of the same functionality. Consequently, it is not possible to enable systematic software patching at the scale at which it might be necessary in course of the long field life, particularly for devices with tight energy consumption constraints (e.g., wearables, implants, many home appliances, etc.) as well as devices with high performance or real-time requirements (e.g., gaming systems, automotives, etc.).

© Springer International Publishing AG, part of Springer Nature 2019
S. Ray et al., *Security Policy in System-on-Chip Designs*,
https://doi.org/10.1007/978-3-319-93464-8_5

The goal of hardware patching is to provide a seamless and disciplined way for hardware implementation of design functionality—in particular security requirements—to be reconfigured in post-silicon and in-field. For this work, we primarily focus on upgrades to security policies. Why is that adequate or even interesting? Because most of the in-field updates to the design *are* in response to security issues. Consequently, it makes sense to focus on that area first when developing technologies for hardware patch.

To use E-IIPS as a basis for hardware patching, we look at three different aspects. First, we look at how we can implement diverse SoC security policies in the architecture in a way that it is possible to adapt, modify, and upgrade these policies in field. Second, we present an automatic security policy mapping flow, providing a streamlined methodology for compiling high-level security policy definitions into the framework. Finally, we provide comprehensive quantitative analysis of energy, performance, and area overhead of the architecture realized with embedded FPGA. We demonstrate the efficacy of the architecture in terms of upgrading with a large number of realistic security policies obtained through our industry contacts. We then compare our evaluation results with alternative implementations.

5.1.1   A Deeper Dive: Limitation of Current Patching Approaches

We consider two scenarios to illustrate the limitations of current practice of security policy implementation.

Case I: Attack on Confidentiality The attack scenario consists of two IPs, namely a trusted crypto engine IP A and an untrusted third party IP B. Ideally, the security policies to set the access control of different IPs are defined at the risk assessment phase. In practice, however, the policies go through continuous refinement through different phases of architecture. In some cases, the process gets extended to early design and implementation activities as new knowledge and constraints keep coming to light. Consequently, the security architects fail to implement a definitive information flow policy to map the IP-specific design constraints at the time of product launch, i.e., time $t=0$. More importantly, the architects connect several IPs in the same network-on-chip (NoC) in SoC with crypto block due to resource constraints and mark the IPs as "safe" to observe part of the keys being exchanged in the communication fabric. Consequently, the untrusted IP B can be exploited by adversaries at a later time t in device life cycle to violate the information flow policy. For instance, the adversary can revoke *Key Obliviousness*, i.e., exploit the malicious IP B to infer cryptographic keys by snooping data from crypto engine on the low-security communication fabric and gain illegal access breaching the confidentiality of the design [6]. The current mitigation approaches usually include the rigorous tasks of overhauling and upgrading the firmware.

Case II: Attack on Integrity An adversary can launch a code injection attack through a malicious or rogue IP by overwriting code segments via Direct Memory Access (DMA). For instance, the attacker can exploit the System Management Interrupt (SMI) handler to write to an address inside System Management RAM (SMRAM), which is basically part of DRAM reserved by BIOS SMI handlers. Based on the vulnerability, the adversary may have control over the address to write, the value being written, or both [44]. Preventive measures to thwart such attacks include identifying memory access requests to DMA-protected regions, and setting up mechanisms to bar DMA requests to all protected accesses.

In the aforementioned scenarios, the current approaches of mitigation fail to offer in-field adaptation of hardware as it involves multiple IPs and requires comprehensive changes in architecture and implementation. The proposed architecture overcomes these key limitations by enabling efficient and secure upgrade of policy implementations after deployment.

5.2   Hardware Patching Infrastructure

Our key observation is that an SoC security policy can be viewed as a sequence of "commands" that specify how to react to certain behavior of IPs and inputs. These commands can be implemented in a separate, centralized IP which communicates with other IPs through a standardized interface. Based on this observation, we need the following three architectural components to enable implementation of diverse policies in a way that is seamlessly configurable in field.

5.2.1   Reconfigurable Security Policy Engine (SPC)

This block acts as the *security brain* of the SoC. It receives communication of relevant security events from the security wrappers in IPs, identifies the security state, and enforces mitigatory actions based on the enforced policies. The key enabler of in-field patching is the centralized location of all security policy implementations within SPC. In particular, patching only requires updating the commands implemented in SPC. This can be seamlessly performed if SPC is in fact implemented on a reconfigurable hardware such as FPGA. The security wrappers do not need update even when the policies are changed in field. Since they are programmable at boot time by SPC, an update only requires reprogramming them to observe and control a potentially different set of signals that correspond to the updated policies.

5.2.2   Smart Security Wrappers

To implement security policies, SPC must *know* some of the internal events of each
IP and communications among IPs. This is achieved through the design of smart
security wrappers, which essentially extend the test (e.g., IEEE 1500 boundary
scan-based wrapper) and debug wrapper (e.g., ARM's CoreSight IP interface) which
are already present for functional verification. The wrappers are programmable, so
that they can be configured to monitor and control different sets of signals. SPC
configures the wrappers during boot for monitoring signals necessary to enforce the
implemented security policies; during execution, SPC identifies the security state of
the system from the monitored signals, and if a policy violation is detected, SPC can
"allow" or "disallow" the event.

Figure 5.1 illustrates an example policy implemented through this architecture.
The policy prohibits access of first 16 (address-wise) internal registers of IP A by
IP B when A is performing a security-critical computation. To enforce the policy,
SPC must know when B attempts to access particular local registers of A as well
as the security state of the computation being performed by A at that instant. When
IP A starts a security-critical computation as indicated by a status flag, its security

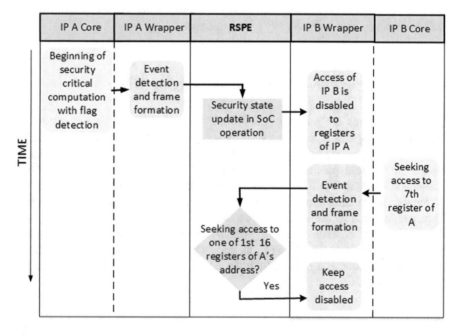

Fig. 5.1 Implementation of representative SoC security policy

wrapper detects the event and communicates it with the SPC. SPC updates the security state of the SoC and disables accesses to all registers of A by B through control logic in B's security wrapper. If B attempts to access a register bit (e.g., register 7) in A's address space to read a configuration value, the security wrapper of B detects this event of interest and informs the SPC. The SPC, determining that the request of B as a violation of the policy, denies corresponding access and maintains the disable status in B's security wrapper.

5.2.3 Integration with Design-for-Debug (DfD) Infrastructure

To obtain controllability and observability over the required signals inside IP blocks, we interface SPC with the on-chip Design-for-Debug (DfD) interface. A primary purpose of DfD is to provide access and control to signals that enable diagnosis of an unanticipated bug in field. Consequently, this interface provides access to an extensive set of observable and controllable signals, which can be repurposed to realize security policies. We architect the security wrappers for each IP by exploiting DfD to identify relevant security-critical events. On detection of such an event, DfD communicates the information to the IP security wrapper that communicates it to the centralized SPC. To enable this functionality without hampering debug usages for the DfD, we implemented IP-level modification of the DfD logic including local debug instrumentation. In particular, noninterference with debug usage requires transmission of security data to SPC via a separate port (instead of repurposing debug trace port and bus), which incorporates additional trigger logic. To address this issue, we employed configuration register interface of corresponding DfD modules that can be configured by SPC through trace port and then security-critical events can be logged into IP-specific security wrappers.

5.3 Implementing SPC: MCU vs. FPGA Trade-Offs

The above observation seems to suggest that the E-IIPS program already provides the ability to perform hardware patching. So, in what ways is it deficient? Unfortunately, programmability of SPC in E-IIPS was achieved by enabling policy implementations as microcode to be executed in the MCU of SPC. This implementation does not satisfy performance and energy costs, which may be particularly relevant for automotive, highway control, and infrastructure systems. In particular, in an MCU implementation, execution consequently incurs the usual overhead of a software implementation. To comprehend that cost, we integrated an initial MCU implementation of SPC on an SoC design and implemented a set of 25 representative security policies. Table 5.1 shows some of the illustrative policies implemented. We discuss overhead results in Sect. 6.4, but the overall summary is that the cost can

Table 5.1 Representative SoC security policies implemented on the proposed architecture

Policy #	Predicate part	Action part	IPs involved
1	User mode and (Mem RD/WR Req. by User | Mem RD/WR Req. by IP A | ...)	RD/WR Addr. within specified range	DLX μP and any other IP with access to system memory
2	Supervisor mode and (Mem RD Req. by User | Mem RD/WR Req. by IP A | ...)	RD Addr. within shared memory range and No WR	DLX μP and any other IP with access to system memory
3	Debug mode and (Trace cells busy | power mgmt. module busy)	No update in power control firmware and no changes in SPI controller Config. Reg.	Power mgmt. module and SPI controller
4	!(Supervisor mode) and (Inst. Mem Update Req. through test access port or SPI controller)	No update of Inst. Mem. allowed	DLX μP
5	Active Crypto mode	No interrupt or Memory Access Req. from the DLX core or any IP is allowed	Crypto module, processor and other IPs access to processor

be almost an order of magnitude or more. This makes the MCU implementation unsuitable as a vehicle for practical hardware patching.

We address this problem by making use of embedded FPGA to implement SPC. In addition to performance and power gains, FPGAs can also streamline the design, e.g., by obviating the need for a large instruction memory when the number of policies is small. Note that for an MCU implementation, even if the number of currently implemented policies is small, one may need to reserve a significant instruction memory to support future extensibility. Furthermore, with embedded FPGAs used significantly in SoC designs particularly for IoT applications (e.g., as custom accelerators), there are robust embedded FPGA deployments in an SoC design with established methodologies that can be exploited for our security architecture. Finally, protection of policy implementation against piracy, cloning, or counterfeiting—a big challenge in IoT devices permitting physical access in-field—is mitigated by existing FPGA bitstream protection mechanisms and can exploit the already mature research on bitstream obfuscation, authentication, and mutable FPGA architectures being designed for security, without requiring enforcement of complex firmware update protocols for policy upgrade and protection. However, in addition to reconfigurability needs, computing devices targeted for IoT applications must also operate under aggressive power/performance constraints. This is particularly true for automotive, highway control, and infrastructure systems that must perform real-time mitigation to any security threats without significantly affecting most of the system functionality, e.g., a connected, driverless car must continue communication with the highway infrastructure and other automobiles on road while monitoring any potential security violation and performing mitigation measures. This means that checks for security policy violation must execute continuously while the system is in operation, contributing significantly to the performance and energy cost. It is imperative that security policies are implemented to be configurable *without compromising the energy and performance constraints*.

5.4 Proposed CAD Framework

5.4.1 Overall Flow and Major Steps

We have developed a CAD framework for synthesizing security policies into an SPC implementation based on embedded FPGA. The framework has the following features: (1) it is amenable to automatic synthesis of arbitrary policies if the policies are described in specific predicate–action format, as shown in Table 5.1, and when the required observable and controllable signals are accessible to the security policy engine; (2) it explores the design space to obtain energy-optimal policy implementations; (3) it allows incremental mapping of policies using partial reconfiguration for field upgrade; and (4) it integrates with the existing FPGA synthesis flow and exploits commercial application mapping tools.

Figure 5.2 illustrates the major steps of the security policy mapping flow into an FPGA fabric. It integrates into conventional FPGA synthesis flow by adding two new steps (shaded in blue) in the front end of the overall flow. In particular, it adds a pre-compilation stage where security policies are parsed and a register-transfer level description is created. The security policies are described as 3-tuple: ⟨**timing, predicate, action**⟩. The timing information indicates either an operating mode or a timing information with respect to global clock. In case of policy 1 and 3 in Table 5.1, the timing information is represented as the operational mode for the DLX processor (user mode) and the operational mode of whole SoC (debug mode), respectively. The predicate information indicates specific conditions based on the IP-internal observable signals or property of the interconnection fabric; the condition is expressed as a Boolean function of multiple observable signals. For the first policy, the predicate is formed by OR-ing two or more signals (e.g., Mem RD/WR Req. by User, Mem RD/WR Req. by IP A). The timing information is AND-ed with the predicate to create a combined condition. The third element is the action to be taken when the joint condition is true. This is done by asserting/de-asserting signals or performing specific checks on a set of variables. For policy 1, the specific action is a check if the RD/WR address is within a range. The policies are parsed and an equivalent Verilog RTL code is generated by accounting for the information on IO ports for all IPs, test/debug ports, and interconnect fabric. The input/output is defined by considering all required observable and timing signals as input and all variables which are controlled as outputs. We represent each policy as an "assign" statement if it does not require using a state element, and as a separate "always" block otherwise.

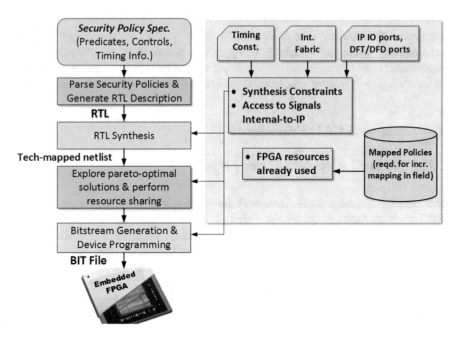

Fig. 5.2 Mapping diverse security policies on embedded FPGA-based SPC

The RTL module representing a set of policies is synthesized using FPGA synthesis tool. By considering the timing and other constraints, we generate a number of pareto-optimal mapping solutions. Resource sharing between several policies, in particular in the implementation of the predicate logic, is explored to optimize the area/energy requirement of the mapping under timing constraint. In case of incremental mapping of new set of policies on the existing policies using the partial mapping solutions provided by the commercial FPGA mapping tools, we use the information on existing resource usage. From the pareto-optimal solutions, we select the best solution in terms of energy or area based on user preferences.

5.4.2   Authentication and Remote Upgrade

To protect the SPC patch installation from attacks by malicious backdoors or Trojans, an authentication mechanism is employed based on secure challenge–response keys. SRAM Physically Unlonable Function (PUF)-based keys are generated at power-up exploiting the intrinsic process variations. The reasons for choosing SRAM PUF include cost efficiency of weak PUF and avoidance of additional circuitry. The power-up key generation prevents any on-chip key storage access control attacks. The PUF-based authentication of the SPC by a verifier at the update unit allows secure remote authentication and in-field patch installation.

5.5   Results and Discussion

We implemented 10 representative security policies of varying complexity (cf. Table 5.1) on our SoC model, including different kinds of access control as well as some instances of information flow, liveness, and secure boot. The entire framework has been functionally validated using ModelSim for typical policy use cases. All area, power, and performance analysis are performed using 32-nm CMOS technology library. We conducted our experiments on Cyclone V FPGA development board.

An estimation of the total number of observable and controllable signals with different design specifications, i.e., *TW* (test wrappers), *SW* (security wrappers), and *DfD* (Design-for-Debug Infrastructure) is provided in Table 5.2. The increase in observability and controllability over signals from various IP blocks is compared among the policy engines with test wrappers (*TW*), smart security wrappers (*SW*), and DfD interface (*DfD*). The column titled *% Increment* represents increase in signal observability and controllability in DfD integrated design compared to security wrapper-based policy engine.

Table 5.3 provides a comparative estimation of the number of arbitrary policies that can be implemented with DfD integrated SPC and other designs. In Table 5.3, *P*, and *A* represents *predicate* and *action*, respectively. We considered the observable signals from multiple IP blocks to determine the possible number of *predicates* and

Table 5.2 Estimation of observable and controllable signals

	IP type	TW	SW	DfD	% increment
Observable	DLX uP	5	547	772	41.13
	AES	5	386	776	101.04
	SPI	n/a	104	161	54.81
	Mem.	5	224	608	171.43
	FFT	5	134	218	62.69
	Total	20	1395	2535	81.72
Controllable	DLX uP	1	142	255	79.58
	AES	1	107	188	75.70
	SPI	n/a	75	144	92.00
	Mem.	1	104	187	79.81
	FFT	1	76	156	105.26
	Total	4	504	930	84.52

Table 5.3 Estimation of number of security policies

Tuple type	TW	SW	DfD	% increment
2P, 1A	570	490,046,760	2,987,015,850	5.10
4P, 1A	14,535	7.91E+13	1.59E+15	19.16
8P, 1A	377,910	1.75E+23	3.89E+25	220.42
8P, 2A	377,910	4.42E+25	1.81E+28	407.94

Table 5.4 Area, performance, power, and energy values for DLX processor core and embedded FPGA-based SPE module and calculated ratios for comparison between two approaches

	Die area (μm^2)	Clock frequency (MHz)	Cycle count (10 policies)	Total latency (μs)	Dynamic power (mW)	Static power (mW)	Total energy (nJ)
DLX μP	0.724	203	210	1.04	14.27	63.48	80.86
FPGA	1.06	138	26	0.189	64.9	20.43	16.13
Ratio	**0.68**	1.47	8.07	**5.49**	0.22	3.11	**5.02**

Bold values indicates obtained ratio between micro-controller and FPGA based implementation for the different metrics

Table 5.5 Results for execution of each policy in FPGA-based SPC

Security policy no.	P1	P2	P3	P4	P5	P6	P7	P8	P9	P10
Energy (nJ)	1.865	1.842	1.851	1.876	1.861	1.839	1.846	1.85	1.868	1.875
Latency (ns)	21.74	14.48	7.24	21.74	14.48	21.74	14.48	14.48	21.74	7.24
Resources (ALMs)	5465	4065	3260	5465	4065	4065	5465	4065	5465	3260

exploited the controllable signals to assert control and set the constraint through the *action* tuple of the policy framework. Table 5.3 shows the maximum limit on the number of security policies in several case scenarios, e.g., 2 *predicates*, 1 *actions*; 4 *predicates*, 2 *actions*; 8 *predicates*, 2 *actions* etc. In every case, the DfD instrumentation results in a higher number of policies compared to other designs.

Table 5.4 shows comparative overhead results for the FPGA implementation over the MCU version. Our testbench exercises 10 of these policies in succession in a specific scenario. For the MCU implementation, between 15 and 20 instructions are involved in the execution of one policy. The estimated dynamic and static powers are based on the signal activities in the representative test bench and standard voltage and thermal models at available 32-nm technology library. The total energy is calculated as the total power (dynamic + static) multiplied by the total latency incurred for the ten policies. For the FPGA implementation, the values of the parameters (die area, latency, total power, and total energy) are calculated as follows. From the Altera Quartus tool, after compilation of the design, the reported number of arithmetic logic modules (both partially and fully utilized ones) is multiplied by the corresponding ALM area to obtain the total die area. The same testbench (as used in MCU version) is utilized to annotate signal activities for dynamic and static power calculation. To mimic the overhead of embedded FPGA, the total reported static power (reported for the whole FPGA chip) is multiplied by the total logic utilization factor to obtain the contribution towards the net leakage power. The FPGA implementation is around 5.02× more energy efficient than the MCU implementation. Furthermore, the MCU implementation takes on average 5.5× more time compared to FPGA to execute these policies. The superior performance and energy efficiency are critical benefits of FPGA implementation since many IoT devices are energy-constrained and often require real-time security protection.

Table 5.5 provides a breakdown of energy consumption per policy. This is done by a testbench that executes each policy in isolation. The energy costs across various policies are close because each policy represents some type of access control

Table 5.6 Comparison of area overhead for the entire SoC

SoC	Org. area (μm^2)	μC design overhead (%)	FPGA design overhead (%)
SoC model	13.1×10^6	21.7	30.74
Apple A6 (APL0598)	96.71×10^6	2.92	4.26
Qualcomm snapdragon 800	118.3×10^6	2.39	3.49

regulation and thereby involves similar computations on the individual IP frames. Besides, all of them incur three cycles (two cycles read of corresponding event frame after buffer flag and one cycle execution to determine security state).

Table 5.6 provides comparison of area overhead for the entire SoC between MCU- and FPGA-based implementation. Even though MCU area is $0.68\times$ of FPGA area, the total area overhead for realistic SoCs is still less than 5%.

5.6 Conclusion

We have presented an architecture and a CAD framework for implementing SoC security policies that accounts for the flexibility and in-field updates required by emerging applications. We have also presented an automatic synthesis framework that enables mapping arbitrary security policies into this framework and integrates with commercial tool flow. High adaptability and capability to implement diverse policies with minimal overhead are distinct features of the architecture. It simultaneously avoids the high performance and energy cost inherent in a software-based policy implementation. For a set of illustrative policies, we showed significant reduction in performance, power, and energy overhead.

5.7 Bibliographic Notes

Early research on security policies looked primarily on software systems and developed analysis frameworks for access control and information flow policies [24]. With modern SoC designs incorporating significant security assets, SoC security policies have become an area of significant research activities. Basak et al. [6] defined an architecture for security policies using dedicated security wrappers. Ray et al. [48] discussed trade-offs between security and debug requirements in SoC designs. Backer et al. [4] analyzed the use of enhanced DfD infrastructure to confirm adherence of software execution to trusted model. Besides, Lee et al. [40] studied low-bandwidth communication of external hardware with the processor via the core debug interface, to monitor information flow. Finally, a recent work has tried to develop languages for formal specification of security policies [42]. Unlike prior work in this area, this paper presents a security architecture and associated CAD flow focused on in-field configurability of security policies.

A preliminary version of this research appeared in the Proceedings of Asia South Pacific Design Automation Conference (ASP-DAC) in 2018 [59]. An overview of the need for patching and informal description of the approach targeted towards a more general audience appeared in IEEE Spectrum [15].

Chapter 6
SoC Security Policy Verification

6.1 Introduction

There are two key motivations for developing a centralized policy engine to implement SoC security policies. One is to provide the ability of hardware patching; the other is to streamline verification of policy implementation. In Chap. 5, we discussed the role of E-IIPS in hardware patching. In this chapter we will discuss its role in verification.

As pointed out in Chap. 1, verification of security policies is nontrivial in the current industrial practice. The hardware logics responsible for protection of various assets are implemented as part of system integration in conflation with various functional and optimization constraints, with little attention paid to ease of verification. In particular, these logics are sprinkled across different IPs. Consequently, *formal* verification of these logics requires discovery, analysis, and comprehension of system invariants that often span across the entire SoC design—an intractable proposition for most industrial SoCs. Furthermore, even *dynamic* (simulation-based) verification is hard, since scenarios exercising security policies involve long, directed execution of corner cases in specific configurations, system execution, and environmental stimuli [60]. Validation of security policies is often left to complex *penetration testing* by human experts and is typically incomplete. Unsurprisingly, security vulnerabilities are discovered in-field, often with disastrous consequences [25, 33].

We take the position that a security *architecture* that facilitates formal verification needs is a feasible solution to the problem. To demonstrate this, we develop a CAD flow on top of E-IIPS for scalable formal verification of SoC security policies. We demonstrate via diverse realistic policy implementations that our approach can result in over $62\times$ speedup (on average) in formal policy verification using state-of-the-art commercial tools. Complex security policies in SoCs that could not be verified at all with traditional implementations become amenable for efficient verification

© Springer International Publishing AG, part of Springer Nature 2019
S. Ray et al., *Security Policy in System-on-Chip Designs*,
https://doi.org/10.1007/978-3-319-93464-8_6

using our framework. Furthermore, our work facilitates short counterexamples in the presence of bugs in policy.

How does E-IIPS help in security policy verification? Recall that policies are implemented in E-IIPS as a state machine defined through a rigorous CAD flow. Policy enforcement entails communication of the policy engine with other IPs in the SoC; this is performed by a standardized protocol. Such centralized policy implementations have an important characteristic to facilitate scalable formal verification: a proof of correctness is typically confined to the policy engine and its interfaces and is oblivious to design invariants in any other IP blocks. Note that all other conditions remaining equal, complexity of formal verification is typically proportional to the size of the design block being analyzed [36]. Consequently, by enabling the target of formal analysis to be confined to a single IP, we enable significant scalability in verification of security policies over traditional distributed implementation. Indeed, we could use off-the-shelf commercial tools without modification to formally verify realistic policy implementations; many of these policies could not be verified at all with traditional approaches.

Our work entails, for the first time to our knowledge, a *formal security verification flow* for security policies that directly influences and exploits architectural support for policy enforcement. Architectural support reduces the complex problem of formally verifying policy implementations to a simpler task of analyzing a single infrastructure IP, thereby providing verification scalability. Second, we demonstrate the efficacy of the framework in identifying the root-causing bugs. Our framework reduces root-causing efforts by providing short counterexamples enclosed to a single IP. This facilitates streamlining the debug flow significantly compared to directed testing and fuzzing approaches used in the current industrial practice. Finally, we develop a comprehensive evaluation of both formal and dynamic aspects of verification on a wide diversity of realistic security policies implemented on an illustrative SoC model. The policies we consider include IP-specific as well as system-level security constraints.

6.2 Problem Analysis

6.2.1 Existing Challenges

Security policies in the current practice are implemented by starting with a baseline architecture which is iteratively refined as follows:

- Use threat modeling to identify potential threats to the current architecture definition.
- Refine the architecture with mitigation strategies covering the threats identified.

The baseline architecture is derived from legacy SoC designs. For each asset, the architect must identify (1) who can access the asset, (2) what kind of access is permitted, and (3) at what points in the system life cycle such access requests can be granted. The current industrial practice for verifying policies includes functional tests, fuzzing tests, and penetration tests. The scale of formal tools is limited to policies involving only one or a few IPs [60].

6.2.2 The Need for Architectural Support

Despite being an integral part of system development flow, the verification requirements of modern industrial designs are barely met by state-of-the-art technologies. Formal methods provide an effective paradigm for security policy validation since they can provide a mathematical guarantee of correctness of the policy enforcement which is unavailable from dynamic (fuzzing, functional simulation, and directed testing) techniques. However, it is imperative to develop architectural support that can make formal analysis of realistic security policies scalable: it is crucial for architectural support that ensures that invariants needed can be enclosed within a small block of logic. Our work facilitates this by basing our CAD flow on top of a centralized framework.

Furthermore, it is important to study the role of policy complexity on verification time. We developed a new metric of policy complexity based on empirical analysis. Figure 6.1 shows the correlation between actual verification time and the proposed metric. The *complexity metric* C_m is defined as follows:

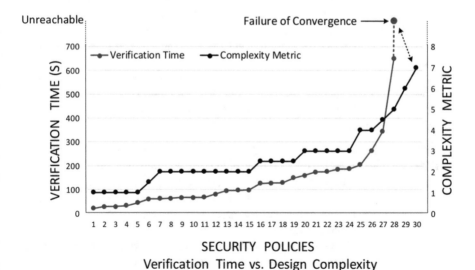

Fig. 6.1 An illustrative example of required verification effort with increasing design complexity and an eventual failure of traditional formal verification approaches

$$C_m = R_c + \sum_{i=1}^{n} \frac{E_i}{2} \left[\frac{O_i}{S_i} + \frac{C_i}{S_i} \right] \qquad (6.1)$$

Here, S_i is the number of security-critical signal(s) involved in a policy, E_i is the number of security event(s) triggered by the policy, O_i and C_i is the number of observable signal(s) and controllable signal(s) of IPs involved in the corresponding policy, respectively. An additional complexity constant R_c is introduced by the SPC in case of the proposed architecture. The soaring verification time (Fig. 6.1) with increasing complexity of policies signifies the limitation of conventional architecture to scale with verification needs.

6.3 Proposed Framework

6.3.1 Architectural Support

Our architecture builds on the centralized policy engine developed in the previous work [6]. Transforming an architecture primarily developed for policy *implementation* (without verification concerns) to support effective formal verification is nontrivial. Here, we list some architectural modifications that are crucial to the verification need.

Event Logging in SPC To facilitate the optimum event detection via centralized architecture, we supplement SPC with augmented event logging capability. The improved event logging is enabled by incorporating configuration register and special purpose registers for storing event trigger, transfer, and related meta data based on event type and requirement. The increment in security events logged by SPC reduces the complexity in security policy verification process by minimizing the reachable trace lengths for verification and violation paths. The centralized implementation of policies and corresponding security properties facilitates the tool to access the required signals in a shorter period of time.

Event Repository in Local DfD A key criteria for the implementation and verification of arbitrary security policies of varying complexity is the detection of a large number of security-critical events in the SoC. System-level security polices of higher complexity often require user-defined triggers and custom interrupts in inter-IP communications. We developed an augmented repository of security-critical events by exploiting the configuration registers in local debug instrumentations. We used on-chip local debug modules with configuration registers and associated logic to map an extended number of security events for system-level policies.

Enhanced Interconnect Fabric We augmented the interconnect fabric of our SoC model for inter-IP communication by establishing a shared memory bus. To facilitate system-level interaction between IPs, we mapped the control registers for each IP to specific addresses of system memory range and utilized the corresponding control signal interfaces to respond to incoming transactions from other functional IPs. In case of incoming interrupts and requests during active computation mode of an IP, a disable signal is instantiated by the policy engine for triggered events. For instance, all requests and interrupts from rest of the IPs are invalidated when AES engine is in crypto mode. Consequently, any unauthorized access requests during crypto mode is logged into configuration registers as potential attempts of violation.

6.3.2 CAD Flow

We automatically synthesize policies into SPC-based architecture. The policies are parsed as *action–predicate* tuples. The principle of pareto-optimality is employed in the synthesis procedure for energy optimum implementation. Figure 6.2 illustrates the design flow. The flow introduces a pre-compilation stage, where security policies are parsed and a register-transfer level description is created for a control state machine that implements the action–predicate tuple; this is integrated with an FPGA synthesis flow to create a reconfigurable policy implementation.

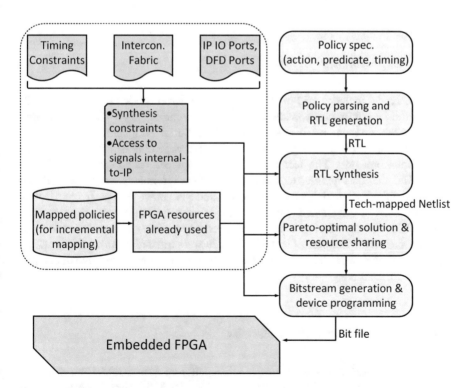

Fig. 6.2 CAD flow for mapping security policies on SPC

Given the above flow, formal property synthesis entails designing of a *monitor state machine* C_p for each policy P. The goal of the monitor state machine is to "watch" C_p and output 1 if p ever makes a deviation from its expected transitions and 0 otherwise. The formal property then reduces to the assertion that C_p never outputs 1. Note that in addition to its use in formal verification, C_p can also act as a runtime monitor for the assertion p: this is relevant in case of a policy p that cannot be completely verified (e.g., if the correctness entails hardware/software co-execution and cannot be established from the hardware alone). In practice, we define C_p by augmenting the RTL design and used primarily for formal verification. If the verification succeeds, the monitor C_p is no longer necessary and can be safely removed: however, if the verification fails or is inconclusive, we keep the augmented RTL and connect the output of C_p to additional routines that can perform mitigatory action if the failure occurs during runtime.

The steps for property mapping are quite straightforward: C_P can be synthesized mechanically from the state machine of P, and is well established in the current industrial practice. However, without centralized SPC, there would be no systematic way for writing these assertions in a traditional SoC design. This further outlines the critical role of architectural support for security verification.

6.4   Results

6.4.1   Experimental Setup

The SoC model includes a 32-bit pipelined DLX microprocessor core (DLX), a 32-KB central system memory, a standard memory controller IP, a 128b AES crypto core, a 128b FFT engine, a clock controller, a Serial Peripheral Interface (SPI) controller, and a power management unit. The IPs were obtained from OpenCores (http://opencores.org). The security policies are mapped on an embedded FPGA-based SPC that acts as the execution engine. For experiments, we developed two versions of the SoC model, a baseline design and SPC-based design. In the baseline SoC, each IP is augmented with standard boundary scan-based wrappers (i.e., IEEE 1500) for detection of local events. Security policies in baseline model are implemented over the constituent IP cores in a distributed manner. In the SPC-integrated model, we enhanced the IPs with smart security wrappers, and developed interface for *DfD* integration.

6.4.2   Formal Verification Results

Our formal verification results use off-the-shelf tool JasperGold [34]. We synthesized assertions in SPC as discussed in Sect. 6.3.2. To compare efficacy, we implemented the same policies on a baseline SoC, paying specific attention to traditional performance optimization to reflect the current state of practice; assertions were also developed for this model. An Intel® Core™ i5-3427U CPU (1.8 GHz) with 8 GB memory is used to run verification on a Linux server.

System-Level Security Policies We implemented 7 (P1–P7) system-level security policies of varying complexity in our SoC model (cf. Table 6.1). Table 6.2 summarizes the verification results for the system-level policies implemented in baseline and SPC-based design. All proven assertions in the table are "Infinite" bound type meaning that the proofs are exhaustive and expected to hold true under all circumstances. We chose multiengine environment with multi-property settings for optimum exploitation of the tool.

The increase in verification time with policy complexity is evident in the results presented in Table 6.2. For instance, the interrupt handling policy (P#3) of AES during crypto mode requires initiation of all possible incoming transactions coming from each of the IPs. The higher verification time, in this case, for DLX interrupts can be attributed to an increased number of security event association. Note that for illustrative system-level policies, our approach reduces verification time by a maximum of 62× compared to baseline implementation.

IP-Specific Security Policies We implemented eight representative IP-specific policies. Table 6.3 summarizes the verification report provided by the tool for baseline and proposed design. The verification of IP-specific policies requires less effort in both baseline and SPC-based design due to the ease of observing and controlling involved signals, with corresponding low verification time.

6.4.3 Scalability Analysis

To demonstrate the scalability of our approach, we consider a case study of boot integrity check policies. These policies verify the trustworthiness of the system at power-on. The implementation of these policies in our model mandates checks for AES crypto engine's data path along with system boot processes (including power-on-self-tests, firmware integrity check, and peripheral core integrity check). Table 6.4 summarizes the verification report. For the baseline design, the verification engines of the tool failed to reach convergence for integrity check policies with multiple engine modes. However, the formal proof is completed in the SPC-based implementation where the state space explosion phenomenon is avoided through the centralized implementation. This suggests that with architectural support, it is possible to address scalability limitations in verification and potentially formally verify complex system-level security policies.

Table 6.1 System-level security policies implemented on the proposed architecture

Policy #	Predicate tuple	Action tuple	Corresponding IPs
1: Read/Write operation of IPs within system memory range in user mode	(Mode: User) and (Memory read/write request by user or any other IP)	Read/write address within specified range	Any IP with access to system memory
2: Read/Write operation of DLX uP to shared memory range in shared memory range	(Mode: Supervisor) and (Memory read/write request by user or any other IP)	Read/write address within shared memory range and No write	Any IP with access to system memory
3: Interrupts (e.g., reset, immediate result, change of key, etc.) from all IPs are prohibited during active crypto mode	(Mode: Active crypto) and (Access request by user or any other IP)	No interrupt or memory access request from any IP is allowed	Crypto module and any other IP with access to crypto core
4: Read/Write access of IPs to round key registers are prohibited during active crypto mode	(Mode: Active crypto) and (Read/write request to round key registers by any IP)	No read/write access to round key registers by any IP is allowed	Crypto module and any other IP with access to crypto core
5: Interrupts by Power management module (clock freq. change, reset, enable, and go) during active computation	(Mode: Active computation) and (Interrupt or access request from PMC module)	No interrupt or access request from PMC module is allowed	PMC module and any other IPs accessible to PMC
6: All IPs' access to interconnect fabric is prohibited during crypto key transfer	(Mode: key in transfer) and (Interrupt or access request by any IP)	No interrupt or access request from IPs to interconnect fabric is allowed	All IPs with privilege to access interconnect fabric
7: Interrupts from all IPs are prohibited during μP core instruction memory update	(Mode: Supervisor) and (μP instruction memory update) and (Access requests)	No access request from any IP is allowed	μP core and any other IP

Table 6.2 Proof of correctness results: system-level security policies

Formal verification results (baseline vs SPC)

Security policies	IP cores involved	Baseline				SPC				Reduction (times)
		JG engine mode	Proof effort	Bound	Avg. time (s)	JG engine mode	Proof effort	Bound	Avg. time (s)	
P #1	DLX up, AES, FFT, SPI	N	1–2	Infinite	27.246	Hp	1	Infinite	1.1435	23.827
P #2	AES, FFT, SPI, PMC	Ht	1	Infinite	64.429	N	1	Infinite	1.387	46.452
P #3	AES, FFT, SPI, PMC	Bm, Hp. I	1–13	Infinite	218.97	Ht, Hp, I	1–4	Infinite	10.824	20.23
P #4	DLX uP, PMC, FFT, SPI	Bm, Hp, Ht,	1–7	Infinite	126.2	N	1–3	Infinite	2.03	62.168
P #5	DLX Up, PMC, FFT, SPI	Bm, D, I.Hp	1–11	Infinite	303.3	I, U, Hp, Ht	2–7	Infinite	21.112	14.366
P #6	DLX uP, PMC, FFT, SPI	Hp, Ht,N	1–3	Infinite	62.958	Ht	1	Infinite	2.7868	22.592
P #7	AES, FFT, SPI, PMC	D, Bm, Hp,N	1	Infinite	142.9	N, Ht	1	Infinite	6.9305	20.619

Table 6.3 Proof of correctness results: IP-specific security policies

Comparative results for IP-specific security policy implementation (baseline vs SPC)

Security policies for DLX up	Baseline			SPC		
	JG engine mode	Bound	Time (s)	JG engine mode	Bound	Time (s)
DLX core instruction memory can only be updated at supervisor mode	Ht	Infinite	1.589	Hp	Infinite	0.695
DLX mode of operation cannot be unknown at startup	N	Infinite	0.615	N	Infinite	0.714
DLX power (high/low) mode of operation check at startup	N	Infinite	0.519	Ht	Infinite	0.936
External read write to DLX to only I/O mapped data memory region	N	Infinite	1.698	Ht	Infinite	1.796
Security policies for AES crypto core	Baseline			SPC		
	JG engine mode	Bound	Time (s)	JG engine mode	Bound	Time (s)
AES key cannot be unknown at startup	Hp	Infinite	0.875	Hp	Infinite	1.537
Key check against previous set (nonce/to prevent replay attack)	Hp	Infinite	1.328	N	Infinite	2.914
In crypto mode, cipher text output interface is disabled	Ht	Infinite	1.537	Hp	Infinite	1.271
AES power (high/low) mode of operation check at startup	Ht	Infinite	0.505	Hp	Infinite	0.713

Table 6.4 Results on scalability analysis on baseline and SPC-based design

Use case scenario: comprehensive policy implementation

	IP cores involved	Baseline				SPC			
		JG engine mode	Result	Bound	Time (s)	JG engine mode	Result	Bound	Time (s)
Policies: boot integrity check	DLX uP,	Ht	Undetermined	1021	31,056.2	Ht	Proven	Infinite	12,984.8
	AES, SPI,	D	Undetermined	872	9246.8	D	Proven	Infinite	4298.8
	FFT, PMC,	I	Undetermined	1543	22,898.9	I	Proven	Infinite	15,998.6
	Sys memory	Hp	Undetermined	987	18,449.7	Hp	Proven	Infinite	11,365.3

6.4.4   Bug Detection

To evaluate the robustness of SPC in bug detection, we injected a set of bugs in the system-level policies. The bugs were inserted with close interaction with industry and are representative of real security bugs detected in industrial environments. Furthermore, the selection aims to cover the spectrum of confidentiality, integrity, and availability requirements of assets in modern SoCs.

Access to Memory Bug We considered a violation scenario where the state machine controlling the memory address register fails to detect the address range breach in the current/overlapping clock cycle. The bug, if goes unmitigated, can lead to unauthorized access of a malicious attacker or restricted IP to secure memory address range. The possible consequences of such breach include a violation of confidentiality and integrity of the assets of secure memory.

Active Crypto Mode Bug In this violation scenario, the status of active crypto signal remains asserted throughout the crypto sequences and is not de-asserted once the operations are finished. The bug can hamper the secure flow of operation as the IPs are blocked from accessing the crypto assets after a crypto operation. The event leads to violation of availability property and consequent unavailability of assets.

Active Computation Mode Bug We considered a violation scenario where the state machine controlling mode of operation of IPs gets stuck in the current state leading to functional failure. The bug is representative of the functional failure of the SoC due to a loss of availability. It directly affects the incoming transactions from power management module and causes stagnation in the flow of execution.

Results Analysis We tested the policies with bugs in a simulation environment. The summary of *functional verification results* is illustrated in Table 6.5. We employed constrained random testing via ModelSim for assertion-based dynamic verification and ran the simulation for more than 24 h. Random functional testing failed to detect any of the bugs in reasonable time ($>$24 h), which highlights the limitation of traditional security policy verification approaches in SoC designs. Table 6.5 also shows the summary of *formal verification results* for 7 system-level security policies. The trace lengths of counterexamples in baseline design are significantly higher than the trace lengths of SPC-based design. With SPC, the engines of formal tool are able to find violation traces with minimal trace attempts leading to reduced trace lengths of counterexamples and improved verification time. Our approach reduces counterexample detection time up to 34$\times$ compared to baseline design.

Table 6.5 Results on bug detection in security policies for baseline and SPC-based design

Functional verification results (baseline vs SPC)

Security policies	IP cores involved	Baseline		SPC		
		Detected bugs	Time (s)	Detected bugs	Time (s)	Reduction (times)
P #1–P#7	DLX Up, AES, FFT, SPI, PMC	N/A	>86,400[a]	N/A	>86,400[a]	N/A

Formal verification results (baseline vs SPC)

Security policies	IP cores involved	Baseline				SPC				
		JG engine mode	Proof effort	Bound	Max time (s)	JG engine mode	Proof effort	Bound	Max time (s)	Reduction (times)
P #1	DLX up, AES, FFT, SPI	N	1–2	Infinite	32.675	Hp	1	Infinite	0.613	34.351
P #2	AES, FFT, SPI, PMC	Ht	1	Infinite	67.145	N	1	Infinite	1.573	15.686
P #3	AES, FFT, SPI, PMC	Bm, Hp. I	1–13	Infinite	345.489	Ht, Hp, I	1–4	Infinite	6.954	26.651
P #4	DLX uP, PMC, FFT, SPI	Bm, Hp, Ht,	1–7	Infinite	186.228	N	1–3	Infinite	1.915	30.155
P #5	DLX Up, PMC, FFT, SPI	Bm, D, l,Hp	1–11	Infinite	263.746	I, U, Hp, Ht	2–7	Infinite	13.573	14.602
P #6	DLX uP, PMC, FFT, SPI	Hp, Ht,N	1–3	Infinite	80.174	Ht	1	Infinite	1.688	25.973
P #7	AES, FFT, SPI, PMC	D, Bm, Hp.N	1	Infinite	185.259	N, Ht	1	Infinite	7.436	13.731

[a]The simulation run time was greater than 24 h, i.e., >86,400 s. No bugs were detected within the time limit

6.5 Conclusion

We have developed an architectural framework to facilitate efficient and scalable formal verification of complex security policies on SoC platforms. Our work, for the first time to our knowledge, marries two highly crucial but typically isolated components of security assurance, architecture, and validation. We show how to develop an architecture that not only enables systematic policy implementation but also scalable analysis and formal verification. The experimental results on realistic SoC models and policies suggest that the approach can reduce verification time for system-level policies by orders of magnitude, help verification of arbitrary policies with varying complexity, and significantly aid the detection of bugs deeply rooted inside the design.

6.6 Bibliographic Notes

Formal methods have a long history going back to several decades. There are several excellent treatises highlighting various aspects of formal methods [9, 29, 37, 56]. Several techniques based on formal methods have been proposed for the verification of security properties [21, 35, 55]. The focus of these works are hardware security issues, i.e., malicious hardware Trojans, side-channel attacks, etc. Their application, however, is limited by the failure to scale with design complexity. Though novel techniques have been proposed for improvement [28], state space explosion is still the major limitation of proving security properties in large SoC designs. Research efforts have been made to address SoC security and verification issues by developing scalable architectural frameworks. Infrastructure IPs are employed to facilitate SoC functional verification, testing, and yield improvement [64, 77]. However, these approaches lack scalable architectural features like centralized or flexible infrastructure IP, standardized interface with IP blocks, and systematic CAD flows.

Chapter 7
SoC Security Policies: Summary and Future Directions

The overall high-level theme of this book revolves around techniques for a systematic, methodical approach towards implementation of system-level security policies in a modern SoC. These policies typically govern the access of security critical assets, sprinkled around in multiple IPs of the SoC, thereby protecting them from underlying confidential, integrity, and availability attacks from the software stack, firmware as well as external interfaces between SoC and the system. The proposed architecture framework revolves around a centralized micro-controlled security policy controller (SPC) which executes the system-level policies stored in its instruction memory (e.g., flash, ROM, etc.) and asserts the necessary security controls. The SPC serves as the single location for analysis, validation, and upgrade of policies, thereby allowing a disciplined, formal approach towards policy implementation. As a result, the existing issues with post-Si validation and on-field upgrades/patch of SoC policies can be substantially alleviated. The SPC communicates with constituent IP blocks through augmented smart security wrapper, which extracts the security-critical internal events of interest from the IP and communicate them with the SPC. These wrappers provide a standard interface to extract security-related information/activity while abstracting out the internal implementation details of the IP. They can be provided by the IP vendors and can be configured by the SPC at boot time to detect subset of events according to particular usage scenarios. We have provided details of the implementation of SPC and IP security wrappers, focusing on the clustering of standard security-critical events according to a few broad category of IPs. Using a representative SoC model in Verilog RTL, the architectural framework is implemented and functionally verified for a set of typical use cases. The hardware overhead of the IP security wrappers and the SPC with respect to a typical modern day SoC has been estimated experimentally to be minimal.

On top of this basic architecture, we developed two critical architectural extensions. First, to reduce the design effort, hardware overhead involved in security wrapper implementation as well as to enable much greater flexibility in on-field

© Springer International Publishing AG, part of Springer Nature 2019
S. Ray et al., *Security Policy in System-on-Chip Designs*,
https://doi.org/10.1007/978-3-319-93464-8_7

upgrades/patches in response to bugs found or changing security requirements on-field and during post-Si validation, we have proposed repurposing the typically resource-rich on-chip debug instrumentation to detect these security-critical events as well as extract new ones for policy upgrade/patch. Taking advantage of the high degree of observability and controllability of IP designs enabled by the local debug trace macrocells (required for post-Si validation and on-field tests), the SPC configures these local DfD at boot time to detect and send the necessary IP events of interest. The potential savings in wrapper area overheads for different IPs due to debug repurposing have been estimated experimentally to be significant as well. Secondly, the security policy framework has been enhanced with the required run-time architecture support to implement IP-Trust aware security policies in order to ensure security and reliability of SoC operation in the presence of inherently untrustworthy constituent third-party IP blocks. Hence, along with the earlier attack models of software and/or SoC–system interfaces, the threat also incorporates Trojan attacks via malicious/covert hardware or firmware logic and/or typical unintentional vulnerability in these third-party IPs. The methodology is based on isolating rogue activity/payload of the Trojan within the corresponding IP itself. Rogue action is detected through verification of correlation between typical high-level IP temporal events via appropriate IP-Trust aware security policies. Any malicious security wrapper can be detected by cross-checking the local DfD associated with the IP. This added architecture support was incorporated into our SoC framework and a set of Trojan use cases analyzed. The associated hardware overhead for this run-time defense against IP-level Trojans in SoC was calculated to be low with respect to modern SoC values.

Finally, we developed CAD framework on top of this architecture to enable patching and verification. Hardware patching requires mapping the policy implementations into an FPGA flow; the result is a framework that enables seamless in-field policy update while still enabling use cases (e.g., with Internet of Things) that require adherence to tight energy or performance constraints. We also showed how formal verification methods can be effectively used on top of the architectural support provided by E-IIPS.

Of course, our framework is still a far cry from an industry-ready infrastructure. Some promising future research directions include analyzing the communication bandwidth and associated power/energy profiles involving communication fabrics of different types between IP security wrappers and SPC in typical usage scenarios. These include analysis with fabrics such as crossbar, bus, or network-on-chips. Effect of these communication patterns on system performance can be analyzed simultaneously. The developed SoC model needs to be enhanced with more IPs of different types and inter-IP interactions can be enhanced to closely represent a large-scale realistic scenario.

References

1. D. Akselrod, A. Ashkenazi, Y. Amon, Platform independent debug port controller architecture with security protection for multi-processor system-on-chip ICs, in *IEEE DATE* (2006)
2. D.M. Ancajas, K. Chakraborty, S. Roy, Fort-NoCs: mitigating the threat of a compromised NoC, in *IEEE DAC* (2014), pp. 1–6
3. J. Backer, R. Karri, Secure design-for-debug for systems-on-chip, in *IEEE ITC* (2015)
4. J. Backer, D. Hely, R. Karri, On enhancing the debug architecture of a system-on-chip (SoC) to detect software attacks, in *IEEE DFTS* (2015)
5. M. Banga, M.S. Hsiao, Trusted RTL: Trojan detection methodology in pre-silicon designs, in *IEEE HOST* (2010), pp. 56–59
6. A. Basak, S. Bhunia, S. Ray, A flexible architecture for systematic implementation of SoC security policies, in *IEEE/ACM International Conference on Computer-Aided Design (ICCAD)* (2015), pp. 536–543
7. A. Basak, S. Bhunia, S. Ray, Exploiting design-for-debug for flexible SoC security architecture, in *DAC* (2016)
8. A. Basak, S. Bhunia, T. Tkacik, S. Ray, Security assurance for system-on-chip designs with untrusted IPs. IEEE Trans. Inf. Forensics Secur. **12**(7), 1515–1528 (2017)
9. J. Bhadra, M.S. Abadir, L. Wang, S. Ray, A survey of hybrid technqiues for functional verification. IEEE Des. Test Comput. **24**(2), 112–122 (2007)
10. S. Bhasin , J.L. Danger, S. Guilley, X.T. Ngo, L. Sauvage, Hardware trojan horses in cryptographic IP cores, in *IEEE Workshop on Fault Diagnosis and Tolerance in Cryptography (FDTC)* (2013), pp. 15–29
11. S. Bhunia, M.S. Hsiao, M. Banga, S. Narasimhan, Hardware Trojan attacks: threat analysis and countermeasures. Proc. IEEE **102**(8), 1229–1247 (2014)
12. L. Bossuet, M. Grand, L. Gaspar, V. Fischer, G. Gogniat, Architectures of flexible symmetric key crypto engines–a survey: from hardware coprocessor to multi-crypto-processor system on chip. ACM Comput. Surv. **45**(4), 41:1–41:32 (2013)
13. R.S. Chakraborty, F. Wolff, S. Paul, C. Papachristou, S. Bhunia, MERO: a statistical approach for hardware Trojan detection, in *Workshop on Cryptographic Hardware and Embedded Systems* (2009)
14. CoreSight on-chip trace & debug architecture, www.arm.com
15. Counterfeit chips on the rise, http://spectrum.ieee.org/computing/
16. A. Das, G. Memik, J. Zambreno, A. Choudhary, Detecting/preventing information leakage on the memory bus due to malicious hardware, in *IEEE DATE* (2010), pp. 861–866
17. F. DaSilva, Y. Zorian, L. Whetsel, K. Arabi, R. Kapur, Overview of the IEEE P1500 Standard, in *IEEE ITC*, pp. 988–997 (2003)

© Springer International Publishing AG, part of Springer Nature 2019
S. Ray et al., *Security Policy in System-on-Chip Designs*,
https://doi.org/10.1007/978-3-319-93464-8

18. L. Davi, A. Dmitrienko, M. Egele, T. Fischer, T. Holz, R. Hund, S. Nurnberger, A.R. Sadeghi, MoCFI: a framework to mitigate control-flow attacks on smartphones, in *NDSS* (2012)
19. H. David, J. Dubeuf, R. Karri, Run-time detection of hardware Trojans: the processor protection unit, in *IEEE ETS* (2013), pp. 1–6
20. Debug specifications, mipi.org
21. S. Drzevitzky, Proof-carrying hardware: runtime formal verification for secure dynamic reconfiguration, in *FPL* (2010)
22. Embedded trace macrocell architecture specification, infocenter.arm.com
23. D. Evans, The internet of things - how the next evolution of the internet is changing everything, in *White Paper. Cisco Internet Business Solutions Group (IBSG)* (2011)
24. J.A. Goguen, J. Meseguer, Security policies and security models, in *Proceedings of 1982 IEEE Symposium on Security and Privacy* (1982), pp. 11–20
25. L. Greenemeier, iPhone hacks annoy AT&T but are unlikely to bruise apple. Scientific American (2007)
26. S.J. Greenwald, Discussion topic: what is the old security paradigm, in *Workshop on New Security Paradigms* (1998), pp. 107–118
27. U. Guin, D. DiMase, M. Tehranipoor, Counterfeit integrated circuits: detection, avoidance, and the challenges ahead. J. Electron. Test. **30**(1), 25–40 (2014)
28. X. Guo, R.G. Dutta, P. Mishra, Y. Jin, Scalable SoC trust verification using integrated theorem proving and model checking, in *HOST* (2016)
29. A. Gupta, Formal hardware verification methods: a survey. Formal Methods Syst. Des. **2**(3), 151–238 (1992)
30. J.T. Haigh, W.D. Young, Extending the non-interference version of MLS for SAT, in *Symposium on Security and Privacy* (1986)
31. M. Hicks, M. Finnicum, S.T. King, M.M.K. Martin, J.M. Smith, Overcoming an untrusted computing base: detecting and removing malicious hardware automatically, in *IEEE Symposium on Security and Privacy (SP)* (2010), pp. 159–72
32. S. Hogg, Software containers: used more frequently than most realize (2014)
33. Homebrew Development Wiki, JTAG-Hack, http://dev360.wikia.com/wiki/JTAG-Hack
34. JasperGold: formal property verification app (2017). www.jasper-da.com/products
35. Y. Jin, Y. Makris, Proof carrying-based information flow tracking for data secrecy protection and hardware trust, in *VTS* (2012)
36. R. Kaivola, S. Pandav, A. Slobodova, C. Taylor, V.A. Frolov, E. Reeber, A. Naik, Replacing testing with formal verification in intel coretm i7 processor execution engine validation, in *CAV* (2017)
37. C. Kern, M. Greenstreet, Formal verification in hardware design: a survey. ACM Trans. Des. Autom. Electron. Syst. **4**(2), 123–193 (1999)
38. S.T. King, J. Tucek, A. Cozzie, C. Grier, W. Jiang, Y. Zhou, Designing and implementing malicious hardware, in *Proceedings of the 1st Usenix Workshop on Large-Scale Exploits and Emergent Threats (LEET)* (2008)
39. S. Krstic, J. Yang, D.W. Palmer, R.B. Osborne, E. Talmor, Security of SoC firmware load protocol, in *IEEE HOST* (2014)
40. J. Lee, I. Heo, Y. Lee, Y. Paek, Efficient dynamic information flow tracking on a processor with core debug interface, in *ACM DAC* (2015)
41. G. Lemieux, D. Lewis, Using sparse crossbars within LUT, in *Proceedings of the ACM/SIGDA International Symposium on Field Programmable Gate Arrays (FPGA)* (2001), pp. 59–68
42. X. Li, Sapper: a language for hardwarelevel security policy enforcement, in *Architectural Support for Programming Languages and Operating Systems (ASPLOS)* (2014)
43. C. Liu, J.V. Rajendran, C. Yang, R. Karri, Shielding heterogeneous MPSoCs from untrustworthy 3PIPs through security-driven task scheduling, in *IEEE DFT* (2013), pp. 101–106
44. J. Loucaides, A. Furtak, A new class of vulnerability in SMI handlers of BIOS/UEFI firmware, in *The 15th Annual CanSecWest Conference (CanSecWest 2015)* (2015)
45. E. Love, Y. Jin, Y. Makris, Proof-carrying hardware intellectual property: a pathway to trusted module acquisition. IEEE Trans. Inf. Forensics Secur. **7**(1), 25–40 (2011)

46. Microsoft threat modeling & analysis tool version 3.0 (2009)
47. ModelSim - leading simulation and debugging. www.mentor.com
48. A.P.D. Nath, S. Ray, A. Basak, S. Bhunia, System-on-chip security architecture and cad framework for hardware patch, in *ASP DAC* (2018)
49. OpenCores. opencores.com
50. P. Patra, On the cusp of a validation wall. IEEE Des. Test Comput. **24**(2), 193–196 (2007)
51. C.P. Pfleeger, S.L. Pfleeger, *Security in Computing* (Prentice Hall, Upper Saddle River, 2007)
52. J. Porquet, S. Sethuamdhavan, WHISK: an uncore architecture for dynamic information flow tracking in heterogeneous embedded SoCs, in *IEEE (CODES + ISSS)* (2013), pp. 1–9
53. J.V. Rajendran, A.K. Kanuparthi, M. Zahran, S.K. Addepalli, G. Ormazabal, R. Karri, Securing processors against insider attacks: a circuit-microarchitecture co-design approach. IEEE Des. Test Mag. **30**(2), 35–44 (2013)
54. J. Rajendran, V. Vedula, R. Karri, Detecting malicious modifications of data in third-party intellectual property cores. in *IEEE DAC* (2015), pp. 1–6
55. M. Rathmair, F. Schupfer, Hardware trojan detection by specifying malicious circuit properties, in *ICEIEC* (2013)
56. S. Ray, *Scalable Techniques for Formal Verification* (Springer, Berlin, 2010)
57. S. Ray, System-on-chip security assurance for IoT devices: cooperation and conflicts, in *IEEE Custom Integrated Circuits Conference (CICC 2017)*, ed. by D. Thelen, K. Tam, A. Pivoccari (2017)
58. S. Ray, J. Bhadra, Security challenges in mobile and IoT systems, in *29th IEEE International System-on-Chip Conference (SOCC 2016)*, ed. by A. Marshall, K. Bhatia, M. Alioto, pp. 356–361 (2016)
59. S. Ray, J. Yang, A. Basak, S. Bhunia, Correctness and security at odds: post-silicon validation of modern SoC designs, in *ACM DAC* (2015)
60. S. Ray, E. Peeters, M. Tehranipoor, S. Bhunia, System-on-chip platform security assurance: architecture and validation. Proc. IEEE **106**(1), 21–37 (2018)
61. J. Rose, V. Betz, FPGA routing architecture: segmentation and buffering to optimize speed and density, in *Proceedings of the ACM/SIGDA International Symposium on Field Programmable Gate Arrays (FPGA)* (1999), pp. 59–68
62. J. Rushby, Noninterference, transitivity, and channel-control security policies, Technical report, SRI, 1992
63. H. Salmani, M. Tehranipoor, Analyzing circuit vulnerability to hardware Trojan insertion at the behavioral level, in *IEEE DFT* (2013), pp. 190–195
64. M. Sastry, I. Schoinas, D. Cermak, Method for enforcing resource access control in computer system, in US Patent 20120079590 A1
65. R. Simha, B. Narahari, J. Zambreno, A. Choudhary, Secure execution with components from untrusted foundries, in *Advanced Networking and Communications Hardware Workshop* (2006), pp. 1–6
66. S. Skorobogatov, C. Woods, Breakthrough silicon scanning discovers backdoor in military chip, in *CHES* (2012), pp. 23–40
67. J. Srivatanakul, J.A. Clark, F. Polac, Effective security requirements analysis: HAZOPs and use cases, in *7th International Conference on Information Security* (2004), pp. 416–427
68. M. Tehranipoor, F. Koushanfar, A survey of hardware trojan taxonomy and detection. IEEE Des. Test Comput. **27**(1), 8–9 (2010)
69. B. Vermeuelen, Design-for-debug to address next-generation SoC debug concerns, in *IEEE ITC* (2007)
70. A. Waksman, S. Sethumadhavan, Tamper evident microprocessors, in *IEEE Symposium on Security and Privacy* (2010), pp. 173–188
71. A. Waksman, S. Sethumadhavan, Silencing hardware backdoors, in *IEEE Symposium on Security and Privacy* (2011), pp. 49–63
72. A. Waksman, M. Suozzo, S. Sethumadhavan, FANCI: identification of stealthy malicious logic using boolean functional analysis, in *Proceedings of ACM CCS* (2013), pp. 697–708

73. X. Wang, Y. Zheng, A. Basak, S. Bhunia, IIPS: infrastructure IP for secure SoC design. IEEE Trans. Comput. **64**, 2226–2238 (2014)
74. W. Wolf, The future of multiprocessor systems-on-chips,in 41*st Design Automation Conference (DAC 2004)* (2004)
75. S. Yerramili, Addressing post-silicon validation challenge: leverage validation and test synergy, in *International Test Conference (ITC 2006)* (2006)
76. X. Zhang, M. Tehranipoor, Case study: detecting hardware Trojans in third-party digital IP cores, in *IEEE HOST* (2011), pp. 67–70
77. Y. Zorian, Embedded memory test and repair: infrastructure IP for SOC yield, in *International Test Conference* (2002), pp. 340–349

Index

© Springer International Publishing AG, part of Springer Nature 2019
S. Ray et al., *Security Policy in System-on-Chip Designs*,
https://doi.org/10.1007/978-3-319-93464-8

Printed in the United States
By Bookmasters